Eloisa Maieski Antunes

Brazil's borders

Eloisa Maieski Antunes

Brazil's borders

the transition of the border strip paradigm in the context of economic integration

ScienciaScripts

Table of contents:

Chapter 1   6

Part 1   15

Chapter 2   15

Part 2   23

Chapter 3   23

Chapter 4   35

Part 3   58

Chapter 5   58

Chapter 6   70

Chapter 7   95

I dedicate this book to my husband **Julimar Luiz Pereira** who brings longevity, prosperity and fun to my life.

Foreword

The scalar gradient is a peculiar feature of border studies. Borders anywhere in the world should not be studied separately within a geographical space because they are immersed in a multi-scalar environment, whether local, regional, state, national or international. Borders also emerge in an environment of multidirectional flows, which are related to social, political and economic issues and directly influence border dynamics by generating flows of blockages and movements.

Brazil is no different. The Brazilian border is one of the longest in the world at 16,886 km and borders ten neighboring countries, making it a very peculiar and diverse area in terms of economy, geography, culture and social aspects, despite belonging to the same territory. The factors that contribute to this diversity range from the population gradients of the municipalities to specific economic factors, historical and political contexts that have a particular influence on each stretch of the border.

With just these two paragraphs, I believe that the reader will begin to understand the complexity of border studies, and this work lends **the gaze of economic and urban geography to the analysis of geopolitical transformations on Brazil's borders over the last two centuries, including population, economic and territorial analysis).**

The hypothesis raised is that the formation of economic blocs has brought about a new territorial configuration because the relationship between space and power has changed, giving rise to a new geopolitical arrangement. The argumentative structures for stating the hypotheses are permeated throughout the text and can be followed by the timeline (figure 0.0).

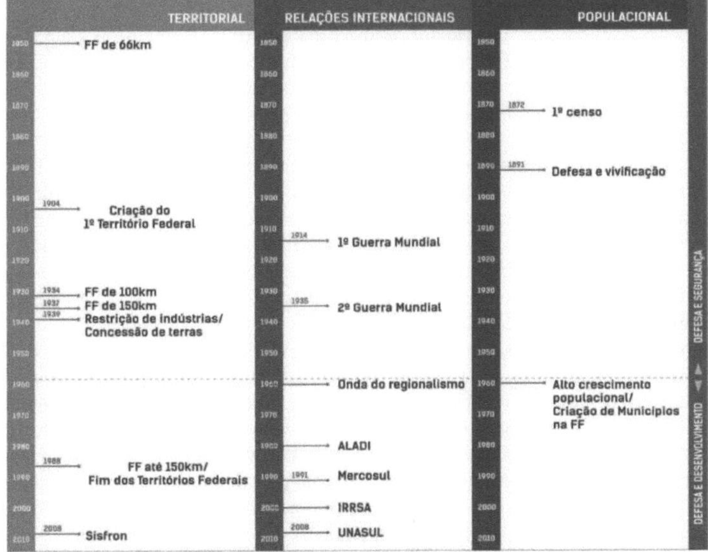

Figure 0.0 : Timeline

Source: Maieski (2018) art by Amanda Mancio (2018)

Between 1934 and 1988, the military saw the border strip (also referred to here as FF) as an area of security and defense. Security is understood as the preservation of sovereignty, the development of political stability and territorial integrity, and defense as the protection of the territory against external threats. Military support in this period was stimulated by the context of the Second World War and the various military dictatorships that were installed in neighboring countries, which influenced border policy in Brazil. Legal restrictions on the border strip were linked to land concessions, means of communication and industries. The Federal Territories were also created, focused on territorial planning on the border, so that the central government had direct control of the border areas considered most vulnerable, in the north, central-west and south.

With the formation of economic blocs, the border strip also began to play new roles, aimed at integrating the countries of South America, mainly because Brazil is a strategic area and neighbors most countries, with the exception of Ecuador and Bolivia. The concept of the border changed when the political and diplomatic context assumed the existence of a Latin American identity, with Brazil in the position of regional leader. In **this way, the border strip, which used to be closed and full of legal restrictions because it was a national defense/security area, came to be seen as a point of strategic contact between Brazil and its neighbors**.

The military paradigm changed with the advent of regionalism, with the promulgation of the 1988 Federal Constitution and with civilian elites pushing for a change in the military perspective in this area. According to Furtado (2013), the border strip went from being an area of national security to one of national defense and development.

Currently, the legislation in force on the border strip is Law No. 6,634, of May 2, 1979, and Decree No. 85, of August 26, 1980, which are important for regulating the activities that can be carried out on the border strip. **Under the paradigm of defense and development, the border strip is now seen as a space of integration, a point of contact with other Latin American countries, as a channel of communication between different cultures, languages and customs. The border strip has thus taken on a new role. Although there are still defense measures, it is now also being treated as a region that needs to be developed economically, especially for foreign trade. This book makes three analyses: territorial, population and international relations to prove the paradigm shift of the border strip from a predominantly defense/security area to an area of development and national defense**.

In order to analyze the **territorial** transformation, this study used data from 1872 to understand the transformations of the border territory from the constitution of the border strip, which was first mentioned in legislation in 1850, to the present day in order to analyze the main characteristics present in the time frame and the territorial planning tools with the width of the border strip and the creation of federal territories. The reader will find that the process of territorial division on the border strip was rapid and intense, especially from 1960 onwards. Between 1940 and 2010, the number of municipalities increased more than sixfold. The military towns set up to defend the border, the combination of economic and political factors with the promotion of the "march to the west", the interest in occupying the borders through vivification (the presence of man), government incentives, the very expansion of the agricultural frontier, the set of infrastructure improvements (roads and railroads) that connected the major urban centers were the factors that led to the multiplication of cities on the border strip.

In the population analysis, we will see that the model used by the military in the last century was that a defended border was an occupied border, but with restrictions on land concessions, industries and means of communication. Currently, the defense model used is through the use of technology via SISFRON - Integrated Border Monitoring (discussed in the last chapter).

In order to analyze **international relations, especially foreign trade,** the chapters that deal with this moment start from the national and international political context that has a direct influence on the place (border). The analysis permeates the border strip from the perspective of economic liberalism, because it has gained a prominent role in the context of economic integration, because it emits its own flows of goods, and is not just a place of passage, and because it is integrated with other regions in the world, not just restricted to neighboring countries.

The end of the Cold War had a variety of meanings for Latin America, the most important of which was undoubtedly the reorientation of foreign policy. The new unipolar world order and the process of re-democratization of the countries are key factors in the new economic issues.

At the beginning of the 1990s, at the height of liberal theories, it was believed that economic interests would predominate in international relations, due to the end of political and military polarization. The effective occupation of national spaces required transforming the classic borders-separation into modern borders-cooperation, in order to make economically viable the development of border regions which, for a long time, had been isolated corners. A fourth element, which becomes more visible in times of financial turmoil, is gender negotiations on a different scale (COSTA, 1999).

The degree of integration between different regions is indicated by the structure of interregional flows of goods and services. The relationship between foreign trade and border municipalities shows that the spatial interactions of the border can be understood on different scales, as they are part of a broad and complex commercial network that maintains a link with the territory.

In the case of Brazil, there is a network of trade at the border and an intensification of flows, but it falls far short of being integrated into an open economy. Brazil has very few companies that venture to export their products, and they don't even account for 1% of Brazilian companies. The economy is still very closed and focused on the domestic market. On the border, despite its geographical advantage of being close to the international border, the same phenomenon occurs. The local economy is still closed to the outside world.

Geographical proximity and tariff agreements are relevant when analyzing the results. Among South American countries, Paraguay and Argentina took turns as Brazil's main trading partners between 1999 and 2013.
MERCOSUR, which highlights the importance of the economic bloc in generating international flows at the border and also the fact that the border strip is the first area of contact with neighboring countries influences the logistics of exporting and importing companies.

The foreign trade flows originating in the border municipalities indicate that in a context of economic integration, the border strip is also capable of emitting its own flows of goods, not just being a place of passage, although trade flows to South America have been losing relative importance as flows are directed to other continents, including Europe, Asia, the Americas and Africa.

It can be seen that the flows generated in the FF are multidirectional. An analysis of the history of foreign trade shows that there have been no variations in the number of countries and that the flows are not directed exclusively to South America. It is therefore possible to say that trade flows are multidirectional in all the years analyzed and are integrated with other regions of the world. The spatial interactions on the border strip indicate that it is capable of producing its own flows with different amplitudes and frequencies because they vary according to the support of the border agents. Thus, the strip is not just a transit corridor. On the border strip, we can see that the flow dynamics follow a national dynamic, but with its own network and link, as will be discussed and presented in the course of this paper.

Geographical proximity is an important factor in analyzing the data. International trade flows to South America have intensified in recent years, but their relative importance has diminished. And trade networks have become stronger with other geographical regions.

The work also studies **SISFRON** - the Integrated Border Monitoring System, which was **included in order to discuss the current defense paradigm,** i.e., since the paradigm shift from defense and security to defense and development is being discussed, the inclusion was necessary for the reader to understand that current geopolitics also maintains defense issues, despite the economic and political environment being favorable to the border economic development model.

Finally, we have tried to organize this book in such a way that you can understand how the border establishes relationships with various sectors of the economy and places in the world, and is therefore not limited to the locality in which it is geographically located. Bearing in mind the themes and arguments to be dealt with, we hope that this book will not necessarily provide ready-made answers on border dynamics, but rather that it can incite you, dear reader, to carry out analyses and reflections that can produce new questions and ideas that help you understand the production of space from the perspective of borders.

In fact, this book is aimed at anyone who wants to know more about Brazil's frontier and the changes that have taken place in this region over the last two centuries.
Happy reading!

Dr. Eloisa Maieski Antunes
Curitiba (Brazil), July 2018.

Chapter 1

1 INTRODUCTION

Brazil's first dividing line was the Treaty of Tordesillas, which delimited the lands that belonged to the Kingdom of Portugal and those of the Kingdom of Spain. Over time, international boundaries were redrawn according to the political and economic interests of the time. Thus, the treaties of Utrecht, Santo Ildefonso and Madrid witnessed the evolution and expansion of Brazilian territory.

The last borders were delimited between the end of the 19th century and the beginning of the 20th century, with the participation of Barao do Rio Branco, who became famous for his application of border policy. He became a famous figure in Brazilian history and left an important territorial legacy for Brazil. The Baron negotiated important treaties with other countries to define international boundaries on the location of borders.[1]

In the fervor of frontier territorial negotiations, for the first time the frontier strip appeared in Brazilian legislation, through Law No. 601 of September 18, 1850, as a geographically delimited area parallel to the international boundaries. There, plots of land could be distributed free of charge by the imperial government to anyone interested in colonizing these distant areas.

The border strip became a geopolitical object par excellence, because it was considered a defense area with the installation of military towns and occupation by man. The[2] focus on the defense and vivification of international boundaries was predominant from the imperial period until the beginning of the Getúlio Vargas government.

The width of the border strip has changed over time. The first law stipulated a strip 66 kilometers

---

[1] The last Brazilian borders were delimited peacefully and diplomatically between the end of the 19th century and the beginning of the 20th century. The work carried out by José Maria da Silva Paranhos Júnior, better known as the Baron of Rio Branco, one of the great personalities in Brazilian history, was fundamental in the negotiation and delimitation of the borders (LESSA, 2012). Ambassador Luiz Felipe de Seixas Corréa comments that for the Baron, geography and history were always his greatest intellectual interests (CORRÉA, 2012).

Rio Branco was the diplomat who, with determination, prudence and knowledge of the facts, led the international arbitration and negotiation processes that contributed decisively to the peaceful establishment of our borders (PATRIOTA, 2012). Before becoming Foreign Minister (1902-1912), he inherited a strategic view of negotiation from his father, the Viscount of Rio Branco. During the adolescence participated, indirectly, in the articulations of the Brazilian empire, especially in the politics applied in the River Plate (CORREA, 2012).

At the end of the 19th century, Brazil was still sewing up the final boundaries, sometimes through international arbitration and sometimes by negotiating territory. According to Correa (2012), the borders had only been definitively fixed with Paraguay in 1872 and with Venezuela in 1859. Territorial conflicts were more marked in the north of the country. Brazil had a territorial dispute with French Guiana, known as the Amapá Question, and a dispute with British Guiana, known as the English Question. In addition, Correa (2012b) states that treaties relating to the fixing of boundaries were signed with the Netherlands (Dutch Guiana) in 1906 and with Colombia in 1907. Boundaries were also set with Peru, after the end of negotiations with Bolivia over the purchase of Acre. The area in dispute over Acre was 442,000 km^2 , which included the 191,000 km^2 incorporated into Brazil by the Treaty of Petrópolis. After five years of negotiations, the boundary treaty was signed on September 8, 1909, giving Brazil 403,000 km2 and Peru around 39,000 km2 (thus reducing Acre to 152,000 km2) (CORREA, 2012b). In the south of Brazil, there was a border dispute with Argentina, known as the Questao de Palmas, which had a geopolitical background according to the prevailing military thinking at the time. A new boundary treaty was signed with Uruguay in 1909 to make up for the excess of rigor committed against the Uruguayans by the 1851 treaty, which had denied them the right of navigation on the Mirim lagoon and the Jaguarao river. In the treaty of October 30, 1909, Rio Branco granted them more than they had demanded: not only free navigation, but also the condominium of the Mirim and Jaguarao lagoons and ownership of some islands (CORREA, 2012b).

[2] The chapter "Legislative tools for territorial planning on Brazil's borders" deals with this classification of military approaches according to the historical moment. Researcher Renata Furtado (2013) cites in her work details on the description of these paradigms.

wide. In 1934, it was extended to 100 kilometers and, later, to 150 kilometers, which it remains to this day.

Between 1934 and 1988, the border strip was seen as a national security zone by the military. The 1934 Constitution included a chapter devoted to national security, and all questions relating to this issue were to be studied and coordinated by the National Security Council. Decree-Law No. 1.164, of March 18, 1939, restricted aspects such as land concessions, means of communication, industrialization and some social and cultural activities on the border strip.

Another legislative tool used in territorial planning was the creation of federal territories on Brazil's borders. At the beginning of the 1940s, between the world wars, the world was going through a period of instability and geopolitical change. This context contributed to the creation of federal territories in regions that the central government considered vulnerable and strategic.

From the 1990s onwards, the international political context interfered with the vision of the state. Under the wave of economic liberalism, many transformations took place on the world political and economic stage, some of which had repercussions in the Belt and Road. One of them is that with the advent of open regionalism, the alternative used to seek financial and economic improvements was the formation of economic blocs in South America, which aimed to standardize customs and non-tariff tariffs. Brazil wanted to expand its export portfolio and the neighboring countries were attractive for carrying out foreign policy more focused on these countries.

The geographical arrangement around economic blocs was also important to be able to compete with countries that had advanced technology and innovation, so it was believed that by building a bloc with common objectives, negotiating on the world stage would theoretically be easier. Various integrationist initiatives emerged, such as: the Andean Community (CAN), the Southern Common Market (MERCOSUR), the Central American Common Market (MCCA), the Caribbean Community (CARICOM), the North American Free Trade Agreement (NAFTA), the Free Trade Area of the Americas (FTAA), among others. The Union of South American Nations (UNASUR) is the most recent and is in the process of being legally and institutionally structured. In general, the blocs in South America have emerged for economic and commercial purposes, although the political aspect is intrinsic to the process.

It's important to note that the integration of South America was an old dream. In the 19th century, Simón Bolivar, known as the Liberator, considered the "prophet of Latin American integration", in an attempt to transform the region into an empire, led armies that liberated five countries in the region from Spanish rule: Colombia, Venezuela, Ecuador, Peru and Bolivia.

Economic integration is a way of "protecting" national companies from the competitiveness of other countries and extending economic borders. Although it is a limited form, the economies within the bloc achieve a more captive consumer market due to the special tax incentives applied. Currently, there are blocs that also aim at the cultural and social integration of peoples; in each one, the stages of deepening are varied and depend on the political will of each state.

Each bloc has its own starting point and specific targets to be met during a given period, as well as its own mechanisms for implementation, control and the resolution of conflicts and disputes.

Thus, the formation of economic blocs brings a new territorial configuration because the relationship between space and power has changed and a new geopolitical arrangement has emerged. In this way, the border strip, which used to be seen by the state as a defense area, has come to play other important roles. Firstly, because the border is the place of contact with neighboring countries, from the moment the government began to form regional economic blocs some border municipalities became strategic because they formed networks of both commercial and social connections linked to other countries and/or other blocs. Secondly, it has gained a role in the context of economic integration by emitting its own flows of goods, not just being a place of passage. Although some authors believe that the foreign trade flows that pass through the border are only transit flows, the municipality's foreign trade statistics indicate that there are flows that cross the border and flows that are emitted at the border. Thirdly, some points along the border trade with other regions in the world, not just with neighboring countries.

The border strip and, more specifically, the border municipalities are inserted in a context of international trade not only with neighbors, but also with more distant countries, due to the generation of trade flows that were encouraged after the liberalism experienced after the war period. Foreign trade thus became another tool for economic development[3] and the border strip took on a new perspective. In addition to being a defense area from the military point of view, the border strip, from 1990 onwards, came to be considered an area of defense and development, giving rise to a new paradigm (defense and development).[4]

The border strip is defined by Law No. 6,634 of May 2, 1979, regulated by Decree No. 85,064 of August 26, 1980, the content of which was ratified by the 1988 Federal Constitution in the second paragraph of Article 20. According to the legislation, the border strip is 150 kilometers wide from the dividing line inwards along the land borders. Although there are still measures to protect against external dangers and for national defense, it is a favorable region for cross-border cooperation and fundamental for affirming regional integration in the Southern Cone. The border strip is next to the other Latin countries and is directly influenced by international economic relations through the economic flows and networks formed in the region.

The Ministry of National Integration and the geographers of the RETIS/UFRJ group carried out a survey of the economic, institutional, cultural and social conditions of the Brazilian border strip and formulated a proposal for the Restructuring of the Border Strip Development Program (PDFF), available on the ministry's *website.*

The program aims to support public policies at various levels of government action. At the same time, the proposal of economic and social indicators can help the private sector make future decisions about investments and actions for citizenship in the border region. The aim of the work is to define a global agenda of guidelines, strategies and action instruments aimed at restructuring the PDFF. The agenda is based on regional economic development and the promotion of citizenship for the inhabitants of the border, at a strategic time for strengthening South American integration (MI, 2005).

The PDFF has adopted a new territorial basis divided into three major arcs: North, Central and South. This "classification" respects the regional, economic, social and cultural differences

---

[3] International economic relations produce a pattern of trade that depends on factors such as labor, productivity, geographical and environmental conditions, the level of technology, capital and the relative use of each factor. Explaining these patterns has been one of the main concerns among economists and economic geographers. This book does not focus on explaining which factors of production produce a pattern of trade on the border strip.

[4] Following the promulgation of the 1988 Federal Constitution, the paradigm applied to the border strip from a military point of view changed from an area of security to one of defense and development (as discussed in Chapter 4). Programs such as Calha Norte were set up to develop and defend the Amazon region, among other functions. One of the most recent and most audacious projects is SISFRON, which aims to monitor the entire border strip, i.e. 27% of the territory, in order to help combat environmental and border crimes. In this vision, the border strip becomes an important object of integration with neighboring countries in order to combat cross-border crimes that occur at certain points along the border.

Political interest in building economic blocs and South American integration also influenced the transformation of the Armed Forces. In 1999, the Ministry of Defense was created, marking the third change in the military paradigm. From a military point of view, the border area became one of defense and development. In 2005, the National Defense Policy document was published and in 2008, the National Defense Strategy (END). These two documents are the most important on the subject because they provide for military cooperation with South American countries in the field of defense.

In 2004, a complementary law was also published on the Army's role on the border strip, authorizing it to act through repressive and preventive actions against cross-border and environmental crimes, along with other bodies. The Army began to direct its practices towards the border strip, mainly in relation to preventive and repressive actions against cross-border and environmental crimes. It can act alone or in coordination with other branches of the Executive, mainly in patrol activities, searches of people and land vehicles, boats and aircraft, and arrests in flagrante delicto (Law No. 97 of June 9, 1999 and Complementary Law No. 117 of September 1, 2004) (FURTADO, 2013).

In particular, the END is organized around three main points, called structuring axes: reorganization of the Armed Forces, reorganization of the national industry and defence material, and composition of the Armed Forces' personnel.

between them. The Southern Arc is made up of the states of Paraná, Santa Catarina and Rio Grande do Sul. The Northern Arc is made up of the states of Amazonas, Roraima, Pará, Amapá and Acre. And finally, the Central Arc comprises Mato Grosso, Rondônia and Mato Grosso do Sul.

Thus, considering the border strip as the region studied in this work, and considering the political and economic moment in which the government chose South America as part of its political identity and economic and commercial relations intensified, creating new forms of organization on the border. In this way, the relationship between territory and capital has become strategic and geographical studies that address this issue are essential for regional development. Given this context, the main question raised in the book is: how has the context of economic integration affected the spatial organization of the border?

The study aims to understand the socio-spatial transformations that have taken place on the border strip before and after the context of economic integration, to understand the dynamics of trade flows and networks generated in border municipalities, highlighting the insertion of some border municipalities in international trade. It will also highlight the cases of special tax regimes for importing and exporting products that are present in some border municipalities.

The analysis of international trade flows is carried out in the 588 municipalities distributed in the three border arcs. The time frame for the demographic and territorial analysis is from 1872 to 2010 and for the analysis of foreign trade flows it is from 1990 to 2014.

1.1 OBJECTIVES

The general objective of this work is:

- Analyze the transition of the border strip from a restricted defense/security area to a defense and development area in the context of economic integration.

The specific objectives of the study are:

1. understand the normative legal tools used for territorial planning on Brazilian borders;
2. analyze the legal and political context for the paradigm shifts in the border strip from 1850 to 1988;
3. analyze the territorial and demographic evolution of border municipalities from 1872 to 2010;
4. to understand the trade flows and networks generated in the border municipalities, highlighting their inclusion in international trade;
5. understand special cases of import/export flows on the Brazilian border.

1.2 HYPOTHESIS

From 1990 onwards, the political context on the border began to change. Under the wave of economic liberalism, many transformations took place on the world political and economic stage, with repercussions on the border strip.

The formation of economic blocs has brought about a new territorial configuration because the relationship between space and power has changed and a new geopolitical arrangement has emerged. In this way, the border strip, which used to be seen more as a defense area, has come to have new roles. Firstly, because the border is the place of contact with neighboring countries, from the moment the government began to form regional economic blocs some border municipalities became strategic because they formed networks of both commercial and social connections, linked to other countries and/or other blocs. Because it has gained a role in the context of economic integration by emitting its own flows of goods, not just being a place of passage. And also because some points of the border strip are integrated with other regions in the world, not just restricted to neighboring countries.

1.3 STRUCTURE OF THE BOOK

The book is divided into 7 chapters. The first consists of an introduction, objectives and hypothesis, briefly addressing the relevance of border studies, with a focus on the geopolitical and economic issues surrounding Brazil's border strip. It also informs the reader of the main geographical concepts covered and the databases used.

The second chapter aims to demonstrate the complexity of the variables involved in border studies, such as the variation in scale (local, state, regional, national and international), the main characteristics of borders and the dynamics of movement and blockage that exist in these areas.

Chapter 3 then describes the two main legal tools used for territorial planning on the border. The border strip and the federal territories were the main instruments for keeping the border territory linked to central power, being used as a security/defense area for the military.

Chapter 4 analyzes the territorial and demographic transformations of border towns from 1872 to 2010. A quantitative study was carried out on the demographic evolution of all available Brazilian censuses and the formation of the border network and the evolution of border towns through the processes of municipal dismemberment, annexation and litigation.

Chapter 5 deals with integration theory and the factors that have led countries to seek new forms of insertion into the global economy. It looks at integration theory and Brazil's economic relations in the light of regionalism in the 90s. It then discusses the changes in territorial policies in South America, highlighting the importance of the Infrastructure Initiative for South America (IIRSA) and the main economic blocs in the Southern Cone. And finally, the history of MERCOSUR. The aim of this chapter is to ground the development paradigm and discuss how international politics can have a direct effect on borders.

Chapter 6 presents the insertion of some border towns in the context of international trade. It analyses the history of import and export flows from border towns, according to the border arcs, the insertion of border towns linked to international trade and the spatialization of exporting and importing towns. Finally, Chapter 7 deals with the defense paradigm and the role of SISFRON as a tool for defending national territory.

1.4 CONCEPTS USED IN THE WORK

The terms used in border studies have generated different interpretations and normative frameworks in the evolution of scientific thought. This section aims to briefly discuss the three main concepts used in the thesis, namely: border strip, border arcs, border municipalities and their variations.

1) The border strip is the first concept to be addressed. Under the 1988 Federal Constitution, art. 20, § 2, the border strip is a geographically delimited area of up to 150 kilometers parallel to Brazil's international boundaries.

2) Border arcs: one of the missions of the Ministry of National Integration, together with other competent ministries and institutions, is to draw up guidelines for the regional development of the border strip. It has adopted a territorial planning criterion based on three geographical zones for the Border Strip Development Program (PDFF) (BRASIL/MIN, 2005). The geographical zones are classified as: northern border arc, central border arc and southern border arc. The division was drawn up by one of the main research groups working on this issue, RETIS, from the Federal University of Rio de Janeiro, made up of researchers, mainly geographers, who developed this new paradigm based on the regional, socio-economic and cultural differences found in each arc.

The Northern Arc is made up of the states of Amazonas, Pará, Amapá, Roraima and Acre. The Central Arc comprises Mato Grosso, Rondônia and Mato Grosso do Sul. And the Southern Arc, Paraná, Santa Catarina and Rio Grande do Sul. Map 1 shows the Brazilian border arcs. In the macro-regional division adopted by the IBGE, Rondonia is part of the Northern region, but in this classification Rondonia belongs to the Central Arc.

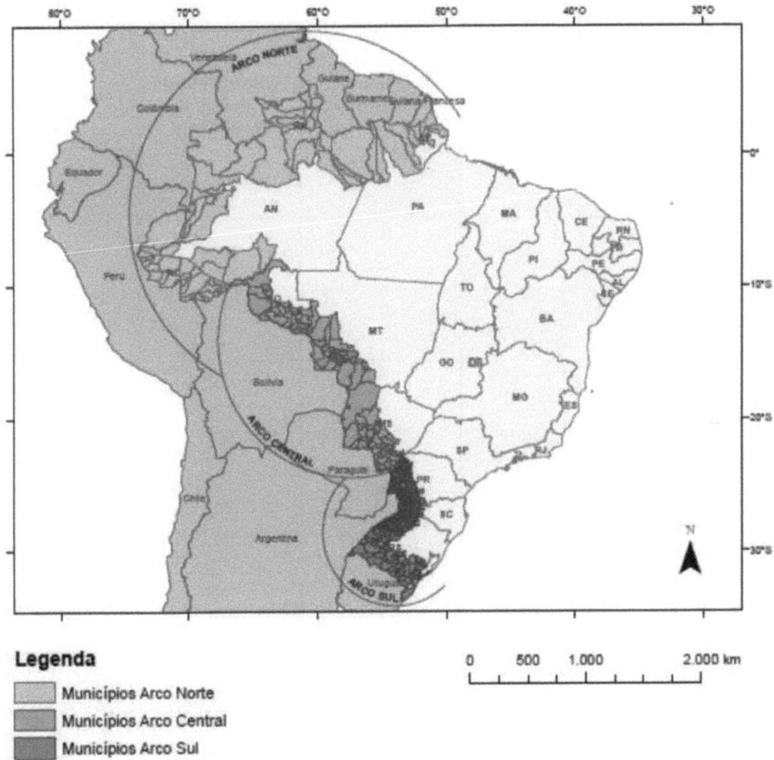

MAP 01 - BORDER ARCS OF BRAZIL'S BORDER STRIP
SOURCE: Ministry of National Integration (2005) - modified by the author

    3) The third concept discussed is that of border municipalities. Border municipalities are those located on the border strip provided for by law. They are classified in three ways:

1. Bordering municipalities: these are those located next to the international boundary. They are not necessarily twin cities.
2. Border municipalities located on the border strip: these are those in which the urban perimeter is located on the border strip (in parts or in its entirety) and does not touch the international boundary.
3. Twin cities: these are border towns whose city center is located right next to the international border.

Table 1 shows the current number of bordering municipalities, those located on the border line and twin cities by state. In the case of Brazil, some of the twin cities came about through the installation of military bases whose function was to provide defense and security for the territory. Over time, the twin cities began to absorb new functions with the implementation of customs and tax bases.

TABLE 01 - DISTRIBUTION OF MUNICIPALITIES
FRONTIERS ACCORDING TO THE CLASSIFICATION ADOPTED BY THE CNM

States	Twin cities	Neighboring municipalities	Border municipalities located on the border strip	Total in the state
Acre	4	13	5	22
Amazonas	1	7	13	21
Roraima	1	7	6	15
Amapá	1	1	6	8
Pará	0	3	2	5
Mato Grosso do Sul	6	7	32	44
Mato Grosso	0	4	24	28
Rondônia	1	8	18	27
Paraná	3	14	122	139
Santa	1	9	72	82
Catherine				
Rio Grande	10	19	168	197
South				
Total	28	92	468	588

Source: National Confederation of Municipalities (CNM, 2008).

In the list of twin cities published by the PDFF in 2005 by the Ministry of National Integration, the municipalities in the state of Mato Grosso do Sul differ from the list published by the National Confederation of Municipalities (CNM). Initially, Corumbá, Porto Murtinho, Bela Vista, Ponta Pora, Paranhos and Coronel Sapucaia were included as twin cities in the PDFF. However, after three years, the CNM published another list which does not classify the municipalities in the same way. The first difference is that Porto Murtinho is considered a border municipality and not a twin city (on a par with Palma Chica - PY). The second difference is the inclusion of Mundo Novo (MS) as a border town, along with Guaíra (PR), from Salto del Guaíra (PY). And the third difference is the exclusion of Coronel Sapucaia from the CNM list.

The criteria for classifying twin cities are being reviewed by the Ministry of National Integration because, although the concept meets a geographical criterion (which is fundamental), an additional criterion is still needed to facilitate the classification of municipalities that wish to be included in this category in order to receive additional benefits (ANTUNES, 2014). This demand began with the regulation of Law No. 12.723, of October 9, 2012, which deals with the installation of *free-shops* in twin cities. After the original list was published, there was a demand from the city halls of neighboring cities to be included in the category of twin cities.

1.5 DATABASES USED

The main databases used were: the Brazilian Institute of Geography and Statistics (IBGE), AliceWeb (Foreign Trade Information Analysis System) and the Ministry of Development, Industry and Foreign Trade (MDIC).

For the development of Chapter 04, the Brazilian censuses from 1872 to 2010 and the territorial division provided by the IBGE were used, so the names of the cities may vary according to the year of the database. Firstly, the *shapes*[5] of each border strip were created with the widths determined (66km - 100km - 150km) in accordance with current legislation. The *shapes* were then superimposed on the political maps of Brazil, reconfigured according to time. The next step was the cartographic analysis of the municipalities that are part or all of the border strip.

For Chapter 6 we used the AliceWeb database - municipalities (export and import), which uses the tax residence of the exporting or importing company as a criterion. It is important to clarify that for a legal entity, the tax residence is equivalent to the registered office or the place where it carries out its activity. The products do not necessarily have to be manufactured in the same municipality.

The analysis of international trade flows was carried out in three ways: 1) analysis of import and export flows at different scales, 2) the relationship between border municipalities and other countries and 3) the centrality of municipalities in the international trade circuit (figure 01).

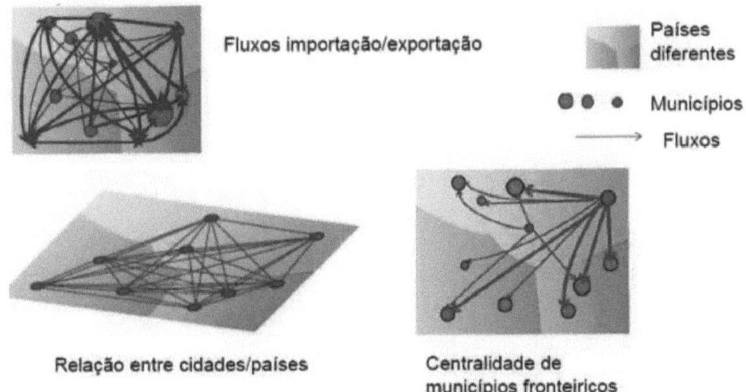

FIGURE 1.1: TYPES OF ANALYSIS CARRIED OUT ON THE INTERNATIONAL TRADE FLOWS OF BORDER MUNICIPALITIES.

Source: Own elaboration

---

[5] A "shape" is a type of digital file that represents a feature or graphic element, whether in point, line or polygon format, and contains a spatial reference (geographical coordinates) of whatever the mapped element is.
A shape is actually a set of several files. Three individual files are required to store the core data that comprises a shapefile. These are the ".shp" file, the ".shx" file and the ".dbf" file. Source: http://www.semace.ce.gov.br/2011/06/shape-definicoes-e-conversao/

For the first stage, the following were analyzed:
- Brazil's import and export flows,
- import and export flows from the border strip,
  -Import and export flows from the border arcs,

Secondly, we analyzed the relationship between the municipalities that are part of the international trade circuit and the countries that have a link.

A third step was to analyze the centrality of the municipalities in the network formed by international trade, by calculating "Betweenness centrality" using specific software. The Betweenness centrality of a node v is given by the expression:

$$g(v) = \sum_{s \neq v \neq t} \frac{\sigma_{st}(v)}{\sigma_{st}}$$

where σ_{st} is the total number of shortest paths from node s to node t and $\sigma_{st}(v)$ is the number of shortest paths that pass through v. It is worth noting that the intermediation of a node scales with the number of node pairs, indicated by the sum of the indices. Therefore, the calculation can be reduced by dividing by the number of node pairs, not including v, so that $g \in [0, 1]$ is divided by $(N-1)(N-2)$ for

directed graphs and by $(N-1)(N-2)/2$ for non- directed graphs.

where N is the number of nodes in the large component. Note that the value scales to the maximum when a node is shared by all the smallest paths. This is usually not the case, and normalization can be performed without losing accuracy.

$$\text{normal}(g(v)) = \frac{g(v) - \min(g)}{\max(g) - \min(g)}$$

which results in: $\max(normal) = 1$

$\min(normal) = 0$

This resizing will always be from a smaller range to a larger range, so there is no loss of precision.

Part 1
Chapter 2

2 The complexity and dynamics of the border
A territory extends beyond borders
and awakens
To adjust the heart to the measure And then together
with your soul the double
We're in different corners of the map Reviewing our
campaign Getting closer to reality To finish a circular
journey. **Beyond the borders. Alfonso Freire**

This chapter aims to briefly present the changing meaning of borders in geographical thinking, the main characteristics of borders and the dynamics that exist in these areas. It addresses the variables involved in border studies, such as the varying scale of the border and the dimensionality of the flows within a complex environment.

2.1 A BRIEF CONCEPTUAL EVOLUTION OF THE BORDER

Discussions on border issues have been prolific and at the same time controversial in the evolution of scientific thought, as they have generated different interpretations and normative frameworks for centuries. The border is created by a political power and is placed before the individual, an individual who belongs to a group, there is a collective memory and identity. It can be interpreted as a mirror of the society that created it, as a barrier and distinction between peoples.

The border is perceived by the inhabitants and used on a daily basis. Its presence produces an effect of barrier and discontinuity, characterizing it as a new element in territorial organization and can also be a field for individual and collective experiences. In one way or another, it occupies a space and ensures a very complex juxtaposition. It is up to the researcher to give meaning to the emerging functions of the objects and territorial fixtures that mark the border context, which can be exclusive, inclusive or associative in relation to non-border factors.

The numerous studies on borders attest to their diversity, the different meanings for the population involved and the implications for public policies. These studies contribute to reflecting on the major economic, social, political, demographic and cultural transformations taking place on the borders between the most diverse countries.

There is a consensus in academic literature that it was with the advent of the modern state that the linear border, precisely delimited and demarcated, became essential, since in order to impose itself the state had to initially lay the foundations of its territorial sovereignty (STEIMAN, 2002). Although the etymological origin of the term *border* was associated with a spontaneous phenomenon of social life, which designated the edge of the inhabited world, over time the systematization of studies in the fields of geography, law, economics and politics brought about a cognitive change in relation to the limit.

For the geographer Michel Foucher, a great thinker on the subject, "borders are elementary, linear spatial structures with the function of geopolitical discontinuity and marking in three registers: real, symbolic and imaginary" (FOUCHER, 2007, p. 38). The notion of reality is the spatial limit of the exercise of sovereignty within the proposed modality; it can be open, closed or ajar. The symbolic dimension refers to the appearance of a political community inscribed in a territory that is treated and identified[6] . The imaginary dimension connotes the relationship with the other, the neighbor, friend or enemy. The migrant or refugee knows very well what it means to cross the "imaginary" line. The border is not a banal functional boundary with a legal and fiscal role (FOUCHER, 1999).

The origin of the word frontier comes from "front", a military term designating the zone of contact with an armed enemy. This sinuous and fluctuating line evolves according to the relationship

---

[6] It is interesting to note that these steps are used by the Brazilian Ministry of Foreign Affairs, as will be discussed later in this article.

of forces present (REITEL, 2004). Originally the term was not applied to a line but to an area. In medieval Europe, the border zone/region was an area, i.e. it was wide and extensive, in order to fulfill the objective of separating undesirable peoples (STEIMAN; MACHADO, 2002).

The idea of a *natural border* emerged in France in the 16th century and was already an indication of the new function of boundaries, related to the foundation of the state's territorial base. The debate that arose between the French and Germans over French sovereignty in Alsace and the Rhine valley (which would be France's "natural border") gave rise to the concept of a border based on the principle that the territorial basis of the state should be linguistic or racial, a position defended by the German humanists (STEIMAN; MACHADO, 2002). The problem surrounding this concept was that the delineation of the natural border was dangerous because it could be strongly contested, as it was not possible to see the land or its map.

From the 17th century onwards, the border progressively became a line at the edge, a limit between two states. It takes on a more political meaning with the construction of nation states, going in the opposite direction to the concept of a natural border because the territorial limit is easier to control than a physical obstacle. The border becomes a line, often artificially drawn, on the land, becoming the object of spatial reorganization. The political border consists of the separation of two territories that are materialized by the existence of a discontinuity, separated by a line. They are two face-to-face political systems, treated as equals, but their functioning, their mode of organization and their legal system are different (REITEL, 2003). The so-called classic spatial borders are those that separate states (WACKERMANN, 2003).

For some authors, there is a difference between a line and a border. Wackermann (2003) mentions that it is advisable to avoid using the notion of a border as a simple boundary. The frontier line is a set of obstacles that results in the reconfiguration of the terrain, the combination of strongholds and the opening of passing fortifications.

The Treaty of Westphalia, signed in 1648, progressively established a new geopolitical order within Europe. The treaty consolidated a set of states, arranged in precise and recognized borders, under which it exercised its power. The fixing and drawing of a border is designed to prevent encroachment on a country's sovereignty and also implies mutual recognition. These rules "guarantee" peace more than war. The Westphalian tradition is associated with the image of the *ligne-frontière* (border line), associated with strong actors such as the state and the representation of the nation. The formation of states is therefore closely linked to the formation of territories and borders.

The border policy adopted by Barao do Rio Branco in the 19th and early 20th centuries saw the process of defining Brazil's borders as a factor of protection and separation, rather than unity. His project did not clearly foresee the establishment of closer ties between Brazil and the South American countries (CORREIA, 2012).[7]

Today the country has a dividing line of 16.886 km, a border population of 11 million inhabitants, a territory of 8,511,965 km^2 and borders with 10 countries[8] . According to Foucher (2007), the contemporary world is structured by 250,000 km of political land borders and 323 interstate borders, and since 1991, more than 28,000 km of new borders have been institutionalized.

With regard to border demarcation, the demarcation of two territories adopted by the Ministry of Foreign Affairs is done by artificial lines, i.e. there is no river or mountain range separating two

---

[7] According to researcher Borba (2013), and also Mattos (1980), the definition of international boundaries had four phases: (i) historical precedents - with a study of the cultural characteristics of the peoples occupying the region with possible attempts to establish the border; (ii) delimitation - through the establishment and ratification of treaties, in an essentially political process, in which the negotiators of the countries decide, in view of the available documentation, how the delimiting line of the territories should be drawn; (iii) demarcation - when the delimiters' intentions are applied to fix the land, river, lake, mountain or other geographical feature chosen as the basis for delimitation, in order to set up landmarks defining the broad lines of the territory's contour; and, finally, (iv) characterization - when there are population occupations along the borders and there is a need to update boundary markers, within the framework established by the demarcators.

[8] Brazil shares borders with 10 countries: Paraguay, Uruguay, Argentina, Suriname, Venezuela, Colombia, Guyana, Peru, Bolivia and Guyana (France), the latter being an overseas department of the French Republic.

territories, given the great difficulty of setting demarcations in geographical accidents, the demarcation lines are geodesic - parallels or meridians. Characterization depends on whether or not the border is populated. In other words, on sparsely populated borders, it is not necessary to set many demarcation marks along the boundary line, but as the border becomes busier, it is necessary to set new marks, making this stage of the process highly technical, since the marks must be increasingly precise (AVEIRO, 2006).

The technique of demarcation by landmarks is ancient. In the 16th century, the Romans used landmarks for the same purpose of demarcating their empire.

FIGURE 02 - BORDER MARKER IN THE CITY OF ACEGUÁ-RS
SOURCE: the author

The MFA's vision is influenced by the Westphalian system, which allowed the emergence of the realist school of international relations, according to which the state was an end in itself, i.e. the state, as the main agent in international relations, should only be accountable to a fictitious national interest, i.e. to itself alone. The Westphalian state marked the officialization of the principles of territoriality and sovereignty, since it became at the same time a government, a territory and a population (AVEIRO, 2006).

From a geopolitical point of view, the border can have a Ratzellian denotation, appearing as a zone of separation and defense between countries. It is also seen as a spatial whole that forms the nation-state, and consequently the state's action ends at the dividing line. The border is considered to be the extreme, the limit of sovereignty, the end of a territory. From the dividing line inwards, the relationship of power and territory belongs solely to the sovereign state. It is a purely political conception due to the authority figure, in this case the government (COELHO, 1992). Therefore, drawing and managing a border are essential acts of applied geopolitics (FOUCHER, 1991).

The author Carneiro Filho (2012) comments that the linear configuration of state boundaries denotes, above all, information, the framing of a political appropriation of space, being one of the geopolitical objects par excellence.

In international law, borders are the product of historical developments. This context can involve occupation as a result of discovery, inheritance law, consideration of the *uti possidetis* principle[9] (RESEK, 2010).

State borders encompass the integrity of the land territory, its internal waters and the sea. Other areas over which the state exercises sovereign rights or jurisdiction are not included in the state's territory (embassies and consulates). International law does not lay down any requirements

---

[9] Through this principle, it came to be understood that a defended border is one where man is present (FURTADO, 2013).

regarding the continuity or discontinuity of territory.

According to Foucher, borders are classically the place where state functions are exercised in peacetime. The legal function reveals the delimitation of a particular sovereignty and the application of a singular national law. Borders are therefore territorial discontinuities based on political markers. In this sense, the institutions established by a political decision, whether concerted or imposed, are governed by a legal text (FOUCHER, 2007).

2.2 THE BORDER IN THE CONTEXT OF INTEGRATION

With the regionalism of the 1990s and the proliferation of economic blocs around the world, which has a strong economic appeal, a new meaning has emerged: the economic frontier. The border here appears as something that should be eliminated, with the political and economic will to expand trade between countries prevailing. The border, as an idea of separation, was not welcome, but the expansion of economic borders to increase economic and political bargaining power was well accepted at this time. Governments and entrepreneurs wanted to expand their businesses and the border in the classical sense was rhetorical. At the time, there was talk of a world without borders.

Moreira comments that for integration to be successful, it needs to be done in three sectors: economic, social and political. The social sector is important because it is impossible to move goods and capital freely without people, without coming into contact with different, often incompatible, cultural models. Therefore, the perception of the population is a fundamental factor if integration is to be successful. This perception is the result of three interacting elements: values, beliefs and information (MOREIRA, 2003).

Integration can also be studied on a micro-scale, which has very particular characteristics. The author Thaise Aveiro makes a distinction between the integration processes underway: a regional process that will govern all integration, mainly in the economic sphere. The other process takes place in a more restricted sphere, that of border integration. These conditions are repeated on all the borders between Brazil and the Southern Cone countries (AVEIRO, 2006).

Border integration is marked by the fact that it is a place of communication and exchange between two distinct territorial domains, as opposed to the international boundary whose defining element is separation. It is a social construction materialized in the relationships between peoples who live the daily life of contact and exchange (SILVA, 2008). Cross-border cooperation is bi-, tri- or multilateral cooperation between public authorities. In Europe, states have largely decentralized their prerogatives, particularly in the economic, social and cultural fields, and territorial collectives are acquiring increasing responsibility for cross-border cooperation. Decentralization laws have increased the responsibility and autonomy of local authorities (GEOCONFLUENCES, 2013). According to Schulz (2002), most cross-border development agreements are based on the economic sphere or are, on average, initiated by economic factors.

It is interesting to note that for many countries the effects of the existence of international boundaries are no longer so important, but rather the effects of the removal of these boundaries or, at least, the removal of the discontinuities that they represented for a long time for economic and social life, for the circulation of ideas, goods and services (STEIMAN; MACHADO, 2002).

These specific, locally-based initiatives have been seen as an essential instrument for intensifying the interrelationships between border communities, a first step towards effective integration. However, even in Europe, where the European Union has led to a certain loss of function of the boundaries between national states more than in other blocs limited to free trade, the action of national governments and supranational organizations has been criticized for its timidity when it comes to border regions. In practice, border or cross-border regions still don't have specific legislation or stimulus projects that are really geared towards them. When action has been taken, it has come from the national governments acting at supranational level in their respective border regions, thus de-characterizing local interaction (STEIMAN, 2002).

Currently, the Brazilian border has a length of 16,886 km, an area of 2,300,000 km^2 and a total of 588 municipalities. In Brazil, the territorial planning of the border areas began to be incorporated from a normative point of view after the construction of three spatial sections for public policies in the study "Proposal for Restructuring the Border Strip Development Program"

(BRASIL/MIN, 2005).

In the Brazilian border strip, a macro-scale was identified, delimited by three border arcs, which in turn were subdivided into analytical meso-scales corresponding to 19 sub-regions, 6 in the Northern arc, 8 in the Central arc and 5 in the Southern arc. In each sub-region it is possible to identify micro-scales that integrate an understanding of the Brazilian border, based on the cross-border scale of the neighboring cities on each side of the border, corresponding to the *twin cities*, which reveal the specificities of the main fixed points that open the way to a dynamic flow. Although there is a small number of *twin cities* on the Brazilian border with South American countries, which highlights the marginality provided by a relative population vacuum and low accessibility of contact, when analyzing the geographical distribution of the *twin cities* it is possible to register a numerical asymmetry between the different territorial arcs.

This heterogeneity in the border areas shows that there is a clear correspondence between the number of border cities and the degree of bilateral or regional cooperation agreements between South American nation states, since almost half of Brazil's *twin cities* are located in the Southern Arc involving border areas with MERCOSUR countries (including the largest of them: Foz do Iguaçu).

Although the original idea of the border appears as a place of separation, border cities have a sense of being a place of accumulation, collection, exchange and maximization of social interactions. The contact zone between two differentiated spaces generates flow dynamics between these two spaces. The interface of a part is more or less wide, from the line to the border zone. A border is not necessarily an interface because it assumes that there is an exchange. Thus, the interface is irrigated, to varying degrees, by flows, more or less intense. Activities, infrastructure and related equipment are often localized (GEOCONFLUENCES, 2013).

According to the work carried out by Bernard Reitel and Patricia Zander (2013), the border has very distinctive characteristics:

a) discontinuity: expresses the difference between two quantitative gradients or differences in legal norms or values. These differences are exploited on both sides of the border where the discontinuity separates contiguous territorial systems. It is manifested by the configurations of the type of attraction/repulsion according to the relative position of one space in relation to another, which generates flows;

b) differentiation: this is the second characteristic factor of the border, and can be economic (differentiation in wages, prices), political (differentiation in tax rules), demographic (available labor force), or even cultural (differences in practices or values). The differential is the relationship between two measurable variables or the difference between legal norms or values. The relationship of reciprocity is not the general rule for cross-border areas, but cases of relationships are possible;

c) domination: this is generally based on the existence of strong gradients (population, labor costs, legal differences, etc.). They actually reveal an ability to organize and structure differences on the border in the best possible way. For example, trade can be unequal between adjacent regions;

d) asymmetry: there are bilateral relationships, but they don't have the same intensity on both sides and don't involve the same characters;

These asymmetries are interesting because they indicate, in addition to differences in the degree of economic development of the countries, different types of regional economy and different dynamics of border settlement. More favorable insertions in the national network-space, geo-environmental conditions unfavorable to settlement, lack of infrastructure linking neighboring agglomerations, political relations between local administrative units and the central government are other factors that influence the urban development of border cities.

The existence or not of symmetries is an important issue. When the systems on both sides of the boundary are analogous, it is likely, as Boggs (1940) *apud* Steiman & Machado (2012) thought, that there will be less tension on the periphery of each, but the existence of symmetry and peace is no guarantee of great interaction. In contrast, House (1980) *apud* Steiman & Machado (2012) postulates that the degree of homogeneity of economic and social conditions on both sides limits the complementarity of exchanges, while great diversity can encourage the development of complementarities and therefore sustain a new cross-border division of labor. For this author, asymmetries and gradient differences are the source of the dynamism of border areas (STEIMAN, MACHADO, 2012).

e) complementarity: as domination is never absolute, there can be some form of complementarity. Each region can be seen as an intermediary, an airlock between two territorial systems;

f) autonomy: cross-border relations are weak, while transnational relations are high. This reveals the existence of a tunnel effect and is seen more frequently in low-density areas. The challenge for these border areas is to develop exchanges that simply don't rely on the difference between national systems.

Given the above, border cities tend to play an integrating role because they are part of an urban network integrated into their territory and because they are close to both sides of the border.

2.3 BORDER FLOWS AND SCALES: A THEORETICAL APPROACH

The border areas formed along Brazil's dividing line with other countries are complex spatial systems and require their own study methodology that reflects the different networks and multi-scales in which these cities are inserted. These systems should not be studied separately within a geographical space because social, political and economic issues directly influence local cross-border dynamics.

Border municipalities are agents of territorial organization because they are part of a territorial fabric that is multi-scalar and multi-dimensional. Multiscale because the border can be analyzed on a local, regional, state, national, cross-border and international scale. It's multidimensional because the flows generated on the border have several directions, depending on the border agent who interferes in the direction of the flows.

In terms of flows, the twin cities in particular (which are also border towns) have both converging vectors - derived from the high potential for integration - and diverging vectors - stemming from the new threats and disputes characteristic of borders. If we analyze, for example, the exchange rate regime and its repercussions on the twin cities, we can see the presence of a divergent economic order. Exchange rate differences generally favor one of the parties.

Daily life on the border also reflects movement. Flows of blockage and movement are present and depend on political factors, most often of a national nature, which directly interfere in the relationship between these flows. As Marques (2007) comments, one vector is related to blocking movement, controlling flows and separation, while the other is related to openness, communication, commercial exchange, information, identity and culture.

The regulation of desirable flows is carried out by the central government. In the Brazilian case, various bodies can propose or standardize regulations that will filter the flows: the Ministry of Foreign Affairs, the Ministry of Development, Industry and Foreign Trade, the Brazilian Federal Revenue Service. International bodies such as the World Trade Organization (WTO), for example, can also interfere with regulations such as diplomatic agreements and treaties, agreements between economic blocs, city halls, state governments, bodies involved in paradiplomatic issues. Thus, each standard will generate specific vectors and flows.

As there is no closed, hermetic border, there are always illegal and clandestine flows on a local scale. Illicit flows across the border in Brazil generate a lot of money. In Foz do Iguaçu, the Brazilian Federal Revenue Service estimates that around US$8 billion worth of goods enter the country annually in the form of contraband/smuggling.

The border is the main filter through which goods and people pass. It is responsible for operationalizing the flows desired by the economic policy of each side. Reitel (2004) comments that customs expresses the porosity of the border and the flow depends on the phases of opening and separating the border, but its main objective is to protect and allow desirable flows to flow.

Customs is responsible for the operational management of foreign trade flows. It is the body responsible for nationalizing cargo entering the country and also registering goods leaving the country. The twin cities that have a customs office on their territory play an important role in internationalization and integration initiatives because they are nodes for redistributing international trade flows.

In Brazil, the institution responsible for customs management, execution, supervision and control is the Receita Federal do Brasil. In addition, this body is responsible for administering internal taxes on foreign trade, repressing smuggling and embezzlement, within the limits of its powers, interpreting, applying and drawing up proposals for the improvement of federal tax and customs legislation, subsidizing the formulation of tax and customs policy and acting in international cooperation and in the negotiation and implementation of international agreements on tax and customs matters (RFB, 2013).

From a geo-economic point of view, the implementation of fiscal/tax and customs bases in the twin cities invites us to rethink borders from the point of view of their porosity, since there is a relative openness to the overflow of positive and negative effects[10] that enhance interstate cooperation or conflict. The border space should not be considered as a spatial system apart from the whole. Customs in the twin cities have a dominant function (opening, filtering and closing), which is determined by political wills located on other scales of power (SENHORAS, 2013).

The twin cities play different roles in the articulation of national flows, and likewise, the production chains are units of economic activities structured in a network that allow us to analyze how different places are involved in the different stages of production and the distribution of their income.

The transportation function is vital for these cities to be included in international production chains, as this sector still complicates the productive development of the inland region of South America due to its distance from the main Atlantic and Pacific ports and the precarious road and rail infrastructure (SILVA, 2008). At this point, the dimension of connectivity between cities takes on special importance, as it no longer depends on the physical distance between them, but on a structure of more or less stable flows, maintained by public and private agents, which reflect not only the characteristics of the past, but also the new forms of insertion into the world market. According to Egler (2001) these flows are proportional to the network of influence that cities exert on their immediate space and what they receive from it, which becomes an element of the city's position in the urban structure.

There is another important issue pending for cross-border regions or border areas. Even if they reach a level of complementarity and effective cooperation, they will need to assert themselves, as they are not just intermediate nodes on the transit routes that link the larger centers to each other. A crucial question for border cities and regions is how to insert themselves into the various transnational networks that cross them, without fatally playing the role of a mere transit point (PRADEAU, 1994 *apud* STEIMAN,MACHADO, 2002).

In the eyes of geopolitics, Brazilian border municipalities have been interpreted as a restricted area of securitization for a long time. In the context of public policies implemented by modern nation states, cities in border areas have acquired a *status of* functional relevance, as fixed points that define boundaries in border securitization. For example, a geopolitical reading can be made of the twin cities of Jaguarao (RS - Brazil) and Rio Branco (Uuruguay).

The author Wanderley Messias (1999) comments that the gauges of the railroad system in Brazil and Uruguay were deliberately different so that there would be no continuity in the route. This reflects a geopolitical context from the past that is still geographically fixed in the city. Figure 03 and 04 show the discontinuity of the rail system between Brazil and Uruguay. The old rail system runs right up to the border and there is no continuity of the railroad tracks. Currently, the Jaguarao town hall has built a walking track for its residents on this site.

---

[10] The negative aspects of the border are problems related to drug trafficking, prostitution rings, smuggling and smuggling.

FIGURE 03 - RAILROAD LINKING URUGUAY'S BORDER WITH MONTEVIDEO
SOURCE: the author
FIGURE 04 - BRAZILIAN SIDE OF THE RAILROAD
SOURCE: the author

The duplication of infrastructure such as parallel highways, airports, power plants, among others, is proof of this competition and a source of wasted resources. This is also a reflection of a conception of border zones as defensive, closed and inward-oriented regions. The construction of these structures is explained by the fact that national developmentism was the dominant ideology, with strong roots in Vargas' nationalist laborism and socialism.

In South America, cities in border areas have undergone an evolutionary transformation in the way they are instrumentalized by the government. Border territorialization was stimulated by the definition of fixed points of limitation and containment in a context based on conflictive geopolitics, which sought to instrumentalize the limitation between internal and external borders, but mainly to territorialize borders through the constitution of cities. At a later stage, the logic developed was essentially geo-economic, although it was also based on geopolitical and geocultural stimuli, and began to foster the socio-spatial formation of border twin cities (CARNEIRO FILHO, 2012).

Nowadays, border towns can be considered hybrid places because they are home to the various contradictions of the system imposed on them, marked by characteristic technical objects such as forts, barracks, customs and inspection posts, and by human actions of diplomatic instrumentalization, diplomatic, carried out by diplomats and presidents, as agents of foreign policy, and paradiplomatic, carried out by sub-national representatives, such as mayors, and by the individual and collective actions of people through a series of convergent or conflictive, legal or illegal flows. In short, all these agents interfere in the border issue, which makes it a complex and multidimensional object.

Part 3
Chapter 3

3 LEGISLATIVE TOOLS FOR TERRITORIAL PLANNING ON BRAZIL'S BORDERS

La frontière, une construction historique évolutive.
Groupe Frontière

This chapter aims to show the two main legal-normative tools used for territorial planning on the border, especially when they were immersed in the paradigm of defense and security. The first was the creation of a border strip, which is a portion of land parallel to the international boundaries, an area legally established by the Federal Government, under the influence of the geopolitical thinking of the time to differentiate it from the rest of the national territory.

The width of the border strip has changed over time due to the war capacity acquired. The first law that established the border strip was Law No.° 601, of September 18, 1850, which stipulated a strip 66 kilometers wide during the empire. Years later, it was extended to 100 kilometers and, later, to 150 kilometers, which remains the case to this day. The chapter analyzes the political and military context of the creation and expansion of the border strip at these different times, highlighting the paradigm shift in the concept of the border strip from national security to national defense and development, as stated by researcher Renata Furtado (2013).

The second tool was the creation of the Federal Territories, which, although now extinct, were used for a number of years as legislative tools for the federal government to have direct control of the border areas considered most vulnerable. The federal territories were created in the north, central-west and south. The federal territories were extinguished after the promulgation of the 1988 Federal Constitution, which is seen as a sign of transition from a paradigm of defense and security to defense and development (mainly economic) due to the context of international relations, which will be discussed in later chapters.

3.1 The current context of Brazil's borders

Brazil's international boundaries have evolved over the course of history. The Treaty of Tordesillas was the first line that separated the colonies that belonged to the Kingdom of Portugal from the colonies of the Kingdom of Spain. As time went by, the boundaries were modified according to political and economic interests; as a consequence, there was a gradual increase in Brazil's territory until it reached the current stage, as discussed in Chapter 3.

The length of the border demarcation line from the north to the south of Brazil measures approximately 16,886 km, making it one of the longest in the world. According to a study carried out by geographer and diplomat Michel Foucher (2007) in his book *L'obsession des frontières,* there are 248,000 km of land borders in the contemporary world, of which 6.33% belong to the Brazilian state. Brazil's longest border is with Bolivia, which is 3,423 km long, followed by Peru with 2,995 km and Venezuela with 2,199 km.

Brazil shares borders with ten countries: Paraguay, Uruguay, Argentina, Suriname, Venezuela, Colombia, Guyana, Peru, Bolivia and Guyana (France), the latter being an overseas department of the French Republic. The Brazil-Guyana border is France's longest land border with a length of 1,605 km, even though it is not a neighboring country. Table 02 shows the length of the borders with each country and the percentage each represents of Brazil's total international boundaries.

The shape and geographical extent of territories, as well as their geodesic position or position relative to the territories of neighboring states, represent factors of political importance that influence the power equation (MATTOS, 1990).

TABLE 02 - EXTENT OF BRAZILIAN BORDERS BY NEIGHBORING COUNTRY AND RELATIVE PERCENTAGE

Ranking	Countries	Total (km)	%

1	Bolivia	3.423	20,27
2	Peru	2.995	17,74
3	Venezuela	2.199	13,02
4	Colombia	1.644	9,74
5	Guyana	1.606	9,51
6	Paraguay	1.366	8,09
7	Argentina	1.261	7,47
8	Uruguay	1.069	6,33
9	Franpa (French Guiana)	730	4,32
10 Total	Suriname	593 16.886	3,51 100

Source: MRE - Ministry of Foreign Affairs (2014).

The border is a geopolitical object par excellence. It is one of the military's functions to be concerned with defending borders in order to assert the sovereignty of the state. Thus, a legal-normative structure for the Brazilian border strip was necessary, drawn up by the military top brass, which will be detailed in the following section.

3.2 Emergence of the border strip

The border strip (BF) is an area legally established by the state to provide different political treatment from the rest of the country. According to Furtado (2013), it is a place for institutional action.

The border strip has a geographical dimension and a legal dimension. The geographical dimension exists because the law geographically delimits an area parallel to the dividing line with neighboring countries. The legal dimension exists because the border strip is a constitutional object and receives differentiated legal treatment. Before being included in the Federal Constitution, the FF was already legally recognized through complementary laws.

The justification used to differentiate the border strip from the rest of the territory is that it is under the rules of either national defense or national security, derived from a military context from different eras[11] . Table 03 shows the main predominant military approaches and each legal

timeframe.
TABLE 03 - LEGAL EVOLUTION OF THE BORDER STRIP AND THE MILITARY APPROACH GIVEN AT EACH HISTORICAL MOMENT

Legal timeframe	Geographical delimitation	Characteristics of the border strip	Focus
1850 (Law no. 601, art. 1)	10 leagues (66 km)	Land area on the empire's borders with foreign countries	Defense linked to vivifying the limits of the empire
Federal Constitution of 1891 (art. 64 and 34)	10 leagues (66 km)	That part of the territory indispensable for the defense of the frontiers as a Union asset	Defense and life
Federal Constitution of 1934 (art. 165)	100 km	100 km strip along borders	National security and national defense
Federal Constitution of 1937 (art. 165 and Law 1.164, of 18/03/1939)	150 km	150 km strip along the borders	National security
Federal Constitution of 1946 (Law No. 2.597/55, art. 180 and art. 34,	150 km	Indispensable area for the country's defense. Internal strip of 150 km from	National security

		width. Portion of wasteland indispensable to the defense of the borders as a Union asset	
II)			
Federal Constitution of 1967 (art. 91, II) 1969 (art. 89)	150 km	Area indispensable to national security. Internal strip 150 km wide, parallel to the dividing line of the national territory.	National security
Federal Constitution of 1988 (art. 20, 2 and Law No. 6.634/79)	Up to 150 km	Fundamental area for the defense of national territory. Border strip up to 150 km wide, along land borders.	National defense and development

Source: Furtado (2013), adapted by Antunes (2014).

The two main approaches are security and national defense. Furtado (2013) defines security as the condition that allows a country to preserve its sovereignty and territorial integrity, to pursue its national interests free from pressures and threats of any kind and to guarantee its citizens the exercise of their constitutional rights and duties. National defense, in turn, is the set of measures and actions of the state, with an emphasis on military expression, to protect the territory, sovereignty and national interests against predominantly external threats, potential or manifest.[13]

For Mattos (2000), national defense places more emphasis on the military aspects of security and, relatedly, on the problems of external aggression. The notion of national security is broader. It includes the preservation of development and internal political stability, and the concept of security is more explicit than that of defense, taking into account internal aggression, embodied in infiltration and ideological subversion, even guerrilla movements.

The expression "border strip" appeared in the 1934 Constitution, but the legal *status of* the FF was recognized in 1850, through Law No. 601 of September 18. In this way, the Border Strip in Brazil was defined for the first time as a delimited geographical area with a special legal regime.

During the Empire, the focus was on defense linked to the vivification of boundaries (FURTADO, 2013). Law No. 601, of September 18, 1850, provided for vacant land in the Empire, reserving a zone of 10 leagues (66 km) contiguous to the boundaries with foreign countries, which could be granted free of charge.

During this period, the government wanted to populate the region and distributed plots free of charge to pioneers (Brazilians or foreigners) who wanted to inhabit it. The land was distributed for agricultural cultivation and animal husbandry. Entrepreneurs who wanted to settle this area had to submit their proposals to the Imperial Government, specifically to the General Department of Public

---

[13] See table by Furtado (2013) on the historical evolution of the size of the border strip.

Lands.

The rules governing the strip of land that separated the Portuguese Colony from the Spanish Colony sought to encourage people to settle in the territory by granting free land to the Crown as a way of defending the borders, applying the principle of *uti possidetis*. This principle led to the understanding that a defended border was one where people were present (FURTADO, 2013).

The first Constitution of the Republic, from 1891, mentions that the federal government was responsible for adopting a convenient regime for the defense of the borders. The delimitation of the portion of the "indispensable territory" for the defense of the borders is not mentioned in the Federal Constitution, but Law No. 601, of 18/09/1850, was still in force, so the first Constitution of the Republic kept the strip under the control of the Union. Therefore, the width of the border strip at the beginning of the Republic was still ten leagues parallel to the international boundaries. Figure 06 shows the 66 km border strip and the municipal seats present. It can be seen that there were few urban centers on the border strip, despite the incentive to occupy these areas.

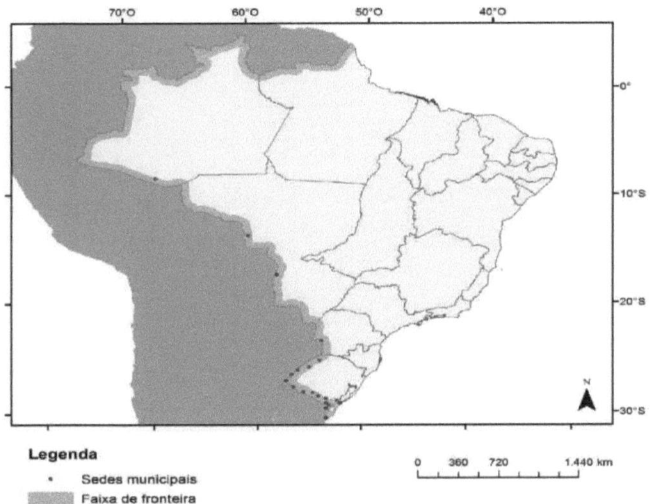

MAP 06 - BRAZIL'S BORDER STRIP AND MUNICIPALITIES
BORDERS IN 1900
SOURCE: prepared by the author

According to Furtado (2013), there were two important events in the country's administrative restructuring during the government of Floriano Peixoto, who granted extraordinary credits for the acquisition of armaments essential for fortifications, especially on the borders of Amazonas and Mato Grosso. And from 1891 onwards, the Ministry of Foreign Affairs was responsible for expediting and dispatching the business and services entrusted to the Ministry of Foreign Affairs, colonization and the service of the colonial nuclei. The geopolitical focus continued to be on "enlivening the frontier"[14] and encouraging the occupation of areas of demographic emptiness.

Despite the efforts to settle people on the frontier, the demographic vacuum prevailed, especially in the north and central-west regions, where access conditions were more difficult. In the south, there were already urban centers on the border, especially on the border with Uruguay. Since the Empire, the border with the state of Rio Grande do Sul was more "lively" than the borders in the center-west and north.

---

[14] An expression widely used in the military environment to characterize the thesis that a defended border is one where man is present. The term is currently found in the National Defense Policy and the National Defense Strategy (FURTADO, 2013).

The width of the border strip changed over time. In 1934, the Federal Constitution extended the FF to 100 km wide (art. 166), as can be seen in Figure 07. The focus given to the border strip changed according to political and military interests. Furtado (2013), in his book "Descobrindo a faixa de fronteira" (Discovering the Border Strip), describes the trajectory of the top military and civilian echelons and the strategies and negotiations to maintain the border strip for years under the national security paradigm.[15]

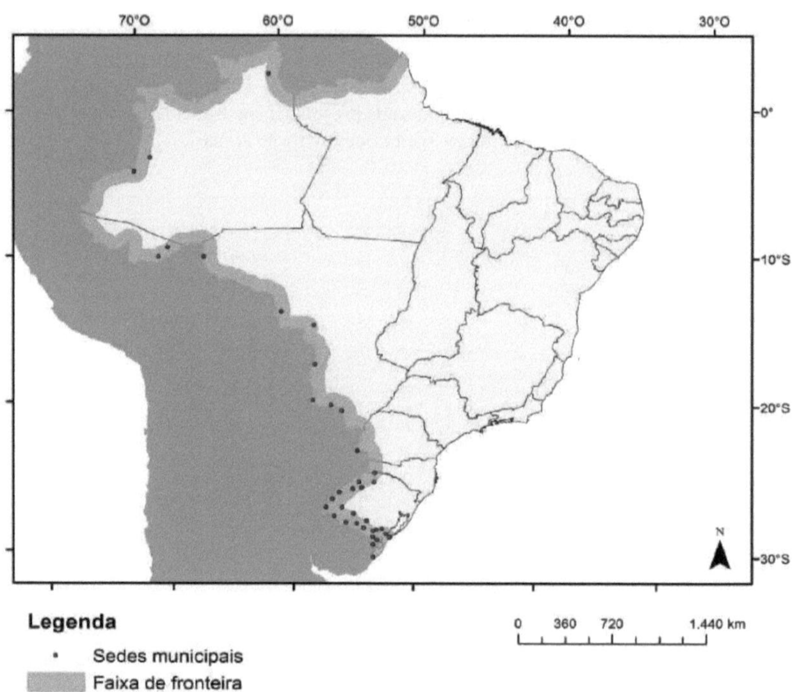

MAP 07 - BORDER STRIP IN BRAZIL AND BORDERING MUNICIPALITIES IN 1933
SOURCE: prepared by the author

The Federal Constitution of 1934 brought two innovations in this regard and, as a consequence, the doctrinal and jurisprudential perplexity about the exact size of the border strip and its repercussions on property began. The first innovation was the extension of the security strip to 100 km and the other was the creation of the concept of security (BARROSO, 1995).

The 1934 Constitution included a chapter devoted to national security, in which all issues relating to this topic were to be studied and coordinated by the National Security Council (CSN) (FURTADO, 2013). Land or communication routes could not be granted without prior authorization from the Council. Furthermore, companies interested in developing this type of activity had to be made up of a majority of national capital and Brazilian workers[16] .

---

[15] Between the Constitution of 1891 and the Constitution of 1934, Decree No. 17,999 of November 29, 1927 was published, providing for the National Defense Council (CDN), which was repealed in 1991. The law on the CDN made no legal changes to the border strip.
[16] Constitution of 1937, art. 165: Within a strip of one hundred and fifty kilometers along the borders, no concession of land or communication routes may be made without the hearing of the Superior Council of

In the 1937 Constitution, the width of the border strip was increased to 150 km along the land boundaries, which has remained the same to this day, although there have been three new constitutions. Figure 08 shows the border strip in 1940.

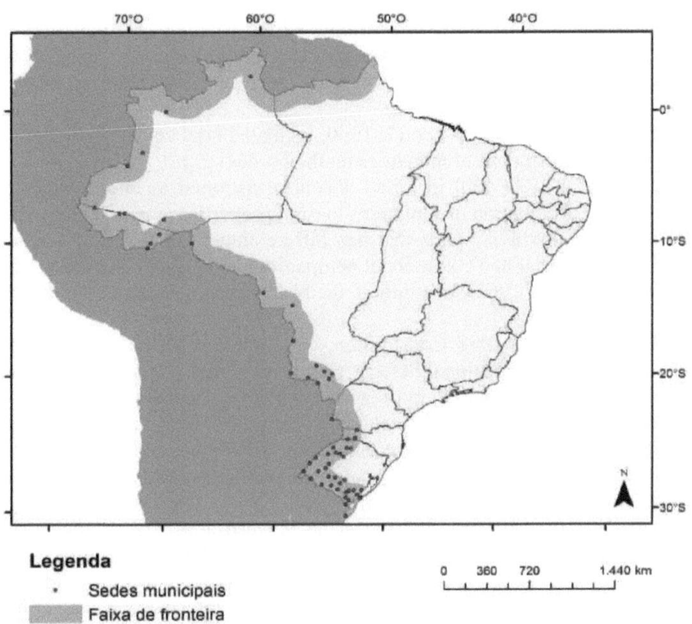

MAP 08 - BRAZIL'S BORDER STRIP AND MUNICIPALITIES
BORDERS IN 1940
SOURCE: prepared by the author

In addition to the 1937 Constitution, new laws relating to the border strip were published. Of particular note is Decree-Law No. 1.164 of March 18, 1939, which governs land concessions and communication routes on the border strip, as well as industries located on the border, helping to establish the idea of a national security area.

The restrictions of Decree-Law No. 1,164 are related to several aspects:

— land concessions: plots could only be granted to native Brazilians (married to native Brazilians) who had the aptitude for agricultural work (art. 4°); land concessions could not exceed two thousand hectares (art. 11);

— means of communication: no concessions relating to means of communication could be granted without the prior authorization of the NSC; concessions had to comply with the requirements imposed by art. 1315;[17]

National Security, and the law shall ensure that in the industries located within the said strip, capital and workers of national origin predominate.

[1715] Art. 13: In assessing the advisability of the concession, from the point of view of the security and defense of the nation, the Council shall also require:

. that the management of the company is entrusted to native Brazilians, or) naturalized for more than ten years;

b) that the administrator has full powers;

. that the company's staff is made up of at least two thirds Brazilians) born or naturalized more than ten years ago;

— industry on the border strip: with regard to industries located on the border, the law required that at least 2/3 of them be made up of Brazilians. Depending on the type of industry, the requirement could be higher, according to art. 13. In addition, the capital of agricultural or industrial companies had to be majority owned by Brazilians (art. 17);

— social and cultural: the law also prohibited the printing or circulation of newspapers, magazines, yearbooks, bulletins and other periodicals in a foreign language. The law provided for the penalty of seizure of copies and closure of the printing house and imprisonment of those responsible for one to three months (art. 18).

Decree-Law No. 1,968 of January 17, 1940, regulated land concessions and communication routes, as well as the establishment of industries on the border strip.

Law No. 6.634 of 1979 (still in force) was also instituted to impose rules on public civil engineering works, the participation of foreigners in rural properties or companies in the FF, land and service concessions and also as a space that has differentiated aid from the federal government. Article 13 stated that industrial and commercial companies had to obtain the necessary authorization from the government.[18] In the 1967 Constitution, the border strip remained an indispensable area for national security.

With the advent of the 1988 Constitution, the paradigm of the border strip as an area of national security became one of national defense and development. According to Furtado (2013), in the period of transition from the paradigm of security to defense, the border strip began to cease to be understood as an indispensable area for national security, becoming a "fundamental" area for the defense of the national territory. This change, introduced by civilian elites in the 1988 Constitution, was decisive in indicating that the FF was no longer just an area of national security, as it had historically been conceived.

The replacement of the word "indispensable defense area" by "fundamental area" in the constitution's wording has a new civilian understanding, since "fundamental to defense would not mean that they were indispensable to defense, that is, this whole dimension could be used if it were not necessary and fundamental, and it could not be used if it were neither necessary nor fundamental to the defense of the national territory" (FURTADO, 2013).[19]

Today, the border strip is considered an area of defense and development. The defense paradigm has once again been adopted because it is more appropriate for the most likely type of current conflict than for external aggression (MATTOS, 2000). And the development paradigm was added because the border strip is now seen as a space for integration, a point of contact with other Latin American countries, as a channel for communication between different cultures, languages and customs.

From a political point of view, the border strip is a favorable region for asserting regional

---

d) that the proportion established in the previous paragraph is observed with reference to the number of employees in the same category;

that a representative of the Federal Government takes part in the administration, with the right *(e)* to freely examine the business and veto any decision, with recourse to the President of the Republic.

[18] Law No. 6.634, of 1979, art. 13: "Industrial and commercial companies that organize themselves, exclusively or not, to operate in the one hundred and fifty (150) kilometer strip along the border of the national territory, including those located in a seaport existing in this strip, must obtain the necessary authorization from the Federal Government, after hearing the National Security Council, through the Special Commission, and may not, under penalty of nullity, enter into operation, nor validly carry out any act, unless they have filed with the Trade Registry, in addition to a certified copy of the act of authorization, the articles of association or statutes, the nominative list of the subscribers, indicating their nationality and the number and nature of the shares held by each one, as well as the respective publication in the Official Gazette of the Union and in the newspapers with the largest circulation in the municipality where they have their headquarters".

[19] The debate and political confrontation between civilians and the military during the 1988 National Constituent Assembly is described in Chapter 5 of the book "Discovering the Border Strip", by Renata Furtado. The book provides information on the political and legislative factors that were decisive for the paradigm shift on the border strip.

integration because the border is the territory that is in permanent contact with neighboring countries. Since the 1990s, the border strip was seen as a defensive, closed and inward-looking region. From the moment the government chose South America as part of its political identity, economic, commercial, cultural and social relations intensified on the border strip, specifically in the border municipalities where the economic flows and networks formed in the region pass through and originate.

The change in perception has not only occurred in Brazil, but also in other South American countries. Furtado (2013) notes that Bolivia, Ecuador, Paraguay, Peru and Uruguay also use a specific width to delimit the border strip, similar to the concept used by the Brazilian Constitution, which involves both defense and development aspects of the border strip (table 04).

TABLE 04 - CHARACTERIZATION OF BORDER STRIPS IN SOME SOUTH AMERICAN COUNTRIES

Country	Characterization	Delimitation	Factor
Argentina	Border zone and security zone of borders	Variable (cartographic)	Security
Bolivia	Area of Border security	50 km 20 km	Security Socio-economic
Colombia	Zones of border	No width indication	Socio-economic
Paraguay	Area of border security	50 km	Security

Peru	Geographical demarcation without the use of specific terminology	50 km	Security and socio-economic
Uruguay	Beech	20 km with Brazil	Socio-economic
Venezuela	Zonade	No indication	Security
	Security	width	

Source: Furtado (2013), adapted by Maieski Antunes (2014). Guyana, French Guiana and Suriname were not cited by the author.

Currently, the legislation in force on the border strip is Law No. 6,634 of May 2, 1979, and Decree No. 85 of August 26, 1980, the most important for regulating the activities that can be carried out in the border strip.

The Ministry of National Integration and geographers from the RETIS/UFRJ group carried out a survey of the economic, institutional, cultural and social conditions of the Brazilian border strip and formulated a proposal to restructure the Border Strip Development Program (PDFF). The PDFF is an important document on a new model for the FF. Currently, border municipalities can be considered hybrid places because they are home to the various contradictions of the system imposed on them, marked by functional objects such as forts, barracks, customs offices, checkpoints, which emerge in a series of converging or conflicting, legal or illegal flows that interfere with the border issue, making it a complex and multidimensional object.

3.3 The emergence of the federal territorial strip

Another legislative tool for territorial planning in the border region was the creation of federal territories. According to Souza et al. (2014), the federal territories were created under the influence of US law and had the function of reproducing the elements of central power, participating in the federative structure for 84 years.

As Porto et al. (2007) point out, the creation of federal territories was configured as "legal prostheses", which are acts drawn up and implemented locally, which impose new rhythms on the primitive environment. Porto (2003) states that the centralization of power over the federal territories was characterized by the high level of participation of this power in local administrative and economic organizations.

The 1891 Constitution did not expressly provide for the possibility of creating federal territories, but the expansion of the rubber industry in Bolivian territory and the subsequent acquisition of Acre resulted in the creation of this type of legal organization (SOUZA *et al.*, 2014). Law No. 1,181, of January 24, 1904, created the Federal Territory of Acre, which remained until 1960. According to the IBGE (2011), this new area was incorporated into the National Territory as a political-administrative unit not linked to any existing state in the Federation, without autonomy and managed directly by the Central Power. Born as a Federal Territory, Acre was only transformed into a federal state in 1962.

At the beginning of the 1940s, the context of the Second World War influenced border policy in Brazil. In 1943, the Federal Government, through Decree-Law No. 5,812, created more federal territories in various places along the border (Figure 09). Decree-Law No. 6,550 of May 31, 1944, stipulated the boundaries and administrative division of the territories. Border policy in Brazil is also linked to demographic policy issues in the Amazon region. Souza (2014) states that the ideology of border protection adopted a closer link with central power, preventing the territory from becoming the object of claims.

The Territory of Amapá is made up of three municipalities: Amapá (40,532,164 km^2 of surface area), Macapá (33,138,343 km^2 of surface area) and Mazagao (46,607,092 km2 of surface area), with Macapá as the capital. The Rio Branco Territory was formed by Boa Vista, the capital, with an area of 95,381,687 km2 and Catrimani (130,664,655 km2). The Guaporé Territory was made up of Porto Velho (23,941,399 km2), Alta Madeira (280,887,183 km2) and Guajará Mirim (89,584,538 km2), with Porto Velho as the capital. The Territory of Ponta Pora was divided into seven municipalities: Pôrto Murtinho (17,758,857 km2 of surface area), Bela Vista (10,051,391 km2), Ponta Pora (25,932,939 km2), Dourados (20,567,752 km2), Miranda (12,930,139 km2), Nioaque (5,137,842 km2) and Maracajú (5,298,028 km2), the last being the capital of the federal territory. The Iguassú Territory was divided into five municipalities: Foz do Iguassú (19,539,539 km2 of surface area), Clevelândia (9,625,567 km2), Iguassú[20] , Mangueirinha (4,296,099 km2) and Xapecó (13,938,795 km2). Foz do Iguassú was the capital.

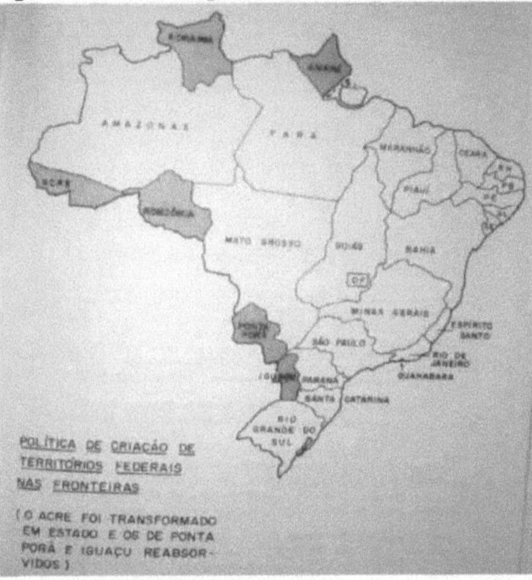

FIGURE 09 - FEDERAL TERRITORIES IN BRAZIL
SOURCE: MATTOS (1980)

The northern region was home to the Federal Territories of Guaporé and Rio Branco. In 1956,

---

[20] The Municipality of Iguassú, according to legislation, refers to the district of Laranjeiras and part of the district of Catanduvas, both in the Municipality of Guarapuava. It was not possible to locate the area at the time using IBGE data.

the name of the Federal Territory of Guaporé was changed to the Federal Territory of Rondônia, which became a unit of the Federation with the Federal Constitution of 1988. The Federal Territory of Rio Branco became the Federal Territory of Roraima, and a unit of the Federation, together with Rondônia.

In the center-west, there was the Federal Territory of Ponta Pora, which was created in 1943 and abolished in 1946. Finally, the Federal Territory of Iguassú was created with part of the territory of Mato Grosso and part of Paraná. However, it was soon extinguished, remaining in existence for only three years and then returning to the original states. According to Temer (1976, *apud* SOUZA *et al.,* 2014), the territories of Ponta Pora and Iguassú were quickly extinguished due to political pressure from the original states.

The 1988 Federal Constitution transformed all the existing Federal Territories into states, but maintained the possibility of the existence of new Federal Territories that would have their governor appointed by the Union, as well as having no representation in the Senate and electing only four federal deputies. Then, in 1988, three more states were added to the Brazilian Federation. These were the states of Amapá, Rondônia - formerly the Territory of Guaporé - and Roraima, formerly the Territory of Rio Branco (IBGE, 2010).

Chapter 4

4 TERRITORIAL AND POPULATIONAL TRANSFORMATIONS ON THE BORDER (1872-2010)

Brazil's border municipalities are highly diverse in terms of their economy, geography, culture and social aspects, despite belonging to the same territory. Some of the factors justifying the diversity found are related to the long length of the border, one of the largest in the world, the population gradients of the municipalities, the specific economic factors, the historical and political moments that have influenced each stretch of the border in a particular way.

Border municipalities are all those that fall within the border strip, either totally or partially. In 1850, the 66 km wide border strip was in force, so municipal perimeters that touched the border strip were considered border municipalities, even if the urban center was not located within the BF.

The aim of this chapter is to analyze the formation and process of division in the border strip from the first census in 1872 to 2010, highlighting the main territorial transformations. It is important to note that the cartographic reading[22] of the municipalities was based on the territorial division made available by the IBGE in each census.

The chapter also includes a study of the spatial distribution of the border population between 1872 and 2010, with data on current population dynamics and the urban hierarchy formed between the municipalities.

The importance of this chapter is to show the reader the territorial transformations, even with the legal restrictions already mentioned in the previous chapter. The intensification of the population on the border is desirable from a military point of view, because it is believed that a defended border is an occupied border. From the point of view of regional integration, human presence on the borders is also positive because it intensifies economic, social and cultural flows and helps to create cross-border integration, although in reality these are only very specific situations. Below we'll take a detailed look at the occupation process from 1872 to 2010.

4.1 Municipal development on the border strip (1872-2010)

Over time, cities emerge in an extremely coherent way. Each city has an original trajectory and specific trends influenced by socio-economic elements (PUMAIN; SAINT JULIEN, 2010) that permeate spatial transformations.

Cities are the result of multiple interactions between authors (neighboring cities, companies, social groups, inhabitants), between their material or symbolic artifacts (equipment, representations) and between events or episodes marked by certain political interventions, economic conjunctures and technological innovations (PUMAIN, 2006).

In Brazil, the urban fabric emerged through an intense and recent process. Between the 1872 census and 1920, the number of municipalities doubled, and between 1920 and 2010 it quadrupled. In less than a century, % of today's Brazilian cities emerged.

The main characteristic of Brazilian urbanization in the 19th century was the concentration of cities near the coast, a legacy of the Portuguese colony which maintained a commercial relationship with European countries. The cities emerged as a result of the economic cycles experienced by Brazil[23] , such as the coastal cities in the northeast as a result of the sugar cane cycle, the *boom of cities in* Minas Gerais was stimulated by the gold cycle and the cities in Sao Paulo and Paraná were greatly influenced by the coffee cycle.

In the first census, in 1872, Brazil had recently proclaimed its independence from Portugal and was seeking international prestige and support. At that time, the country had a large territory in which the population and the main cities were concentrated on the Atlantic coast (Figure 10).

---

[22] For more details on shape construction, see Chapter 01.

[23] The economic cycles discussed are derived from the theory of Furtado (2007) in which he discusses the main economic cycles in Brazil. The work referred to is: FURTADO, C. *Formagao económica do Brasil.* Sao Paulo: Companhia das letras, 2007.

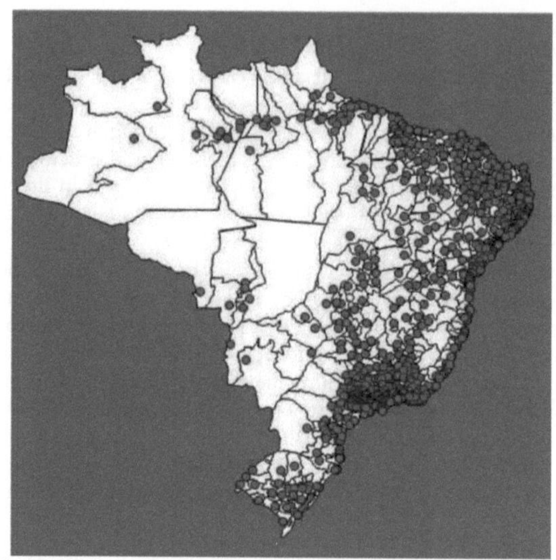

FIGURE 10 - MUNICIPAL SEATS IN 1872
SOURCE: IBGE. Own elaboration

In relation to the border territory[24] of 1872, the main characteristics, present from the imperial era to the present day, are: the territorial area of the municipalities of Arco Norte and Central are derived from units of the federation that already had "giant" territorial areas compared to the municipalities of Arco Sul; the number of municipalities in Arco Sul is greater than the number of municipalities in Arco Norte and Central since 1872.

The border states in imperial times were: Amazonas, Pará, Matto Grosso[25] , Paraná and Rio Grande do Sul, with different dimensions to today, as shown in Table 05. The border states of Santa Catarina, Mato Grosso do Sul, Rondonia, Roraima and Acre did not officially exist. They were created and made official at different times. As the country evolved politically and administratively, these states were included, along with others, to form the national territory.

TABLE 05 - SURFACE AREA OF BORDER STATES IN 1872 AND 2010

States	Surface area in 1872	Area in 2010	Loss of territory
Amazonas	1,851,882,016 km2	1.559.159,148 km2	15%
Pará	1,392,456,634 km2	1.247.954,666 km2	10%

---

[24] Border territory comprises the part of the national territory made up of cities that fall within the border strip.
[25] Matto Grosso was the spelling used at the time to refer to the state of Mato Grosso.

Matto Grosso/ Mato Grosso	1,457,281,503 km2	**903,366.192** km2	38%
Paraná	226,485,210 km2	**199,307.922** km2	12%

SOURCE: IBGE. Own elaboration

The creation of border municipalities was also exponential and rapid, as was urbanization in the rest of Brazil. In the 1872 census there were only 23 municipalities (only 9 with urban headquarters). This census indicated that there were 6 border municipalities that today are classified as twin cities, located in the provinces of Rio Grande do Sul and Matto Grosso (spelling of the time).

Until then, there was no municipality in the northern region of Brazil with a municipal seat located within the border strip.
66 km wide, but some "villas"[26] (small communities that had not yet been emancipated), indigenous villages and military towns, such as Tabatinga, in Amazonas.

Graph 01 shows the relationship between the number of border municipalities and the number of municipalities in the border provinces (which later became states of the federation) and the total number of municipalities in Brazil. In 1872, there were 641 municipalities and 5 border provinces in Brazil. The number of municipalities on the border strip was much smaller than in the rest of Brazil. In proportional terms, in 1872 the total number of border municipalities was only 2.41%, while today this ratio is 5.65%.

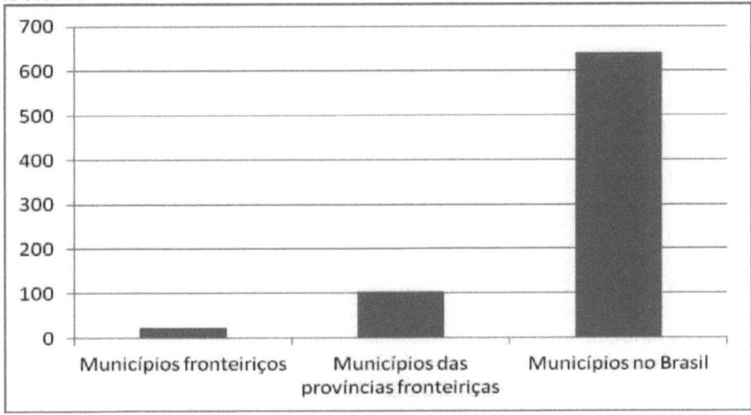

GRAPH 01 - TOTAL NUMBER OF BORDER MUNICIPALITIES, MUNICIPALITIES IN BORDER PROVINCES AND BRAZILIAN MUNICIPALITIES IN 1872
SOURCE: IBGE. Own elaboration

The process of territorial division on the border strip has been rapid and intense, especially since 1960. The number of border municipalities in 1940 was 96, and today, according to the latest census, there are 588 municipalities; thus, between 1940 and 2010, the number of municipalities increased more than sixfold. Graph 02 shows the evolution of the number of border municipalities from 1872 to 2010.

In comparison, the number of border municipalities became more significant from the 1960s

---

[26] There are actors who claim that these villages were already considered municipalities. However, as the research was based on data from the IBGE (Brazilian Institute of Geography and Statistics), the records available on www.ibge.gov.br were taken into account.

onwards, at a time when from a legal-military point of view the border strip was still considered a national security area. The military towns set up to defend the border, the set of economic and political factors with the promotion of the "march to the west", the interest in occupying the borders through vivification (the presence of man), government incentives, the very "expansion of the agricultural frontier", the set of infrastructure improvements (roads and railroads) that connected them to the major urban centers, were the factors that led to the multiplication of towns on the border strip.

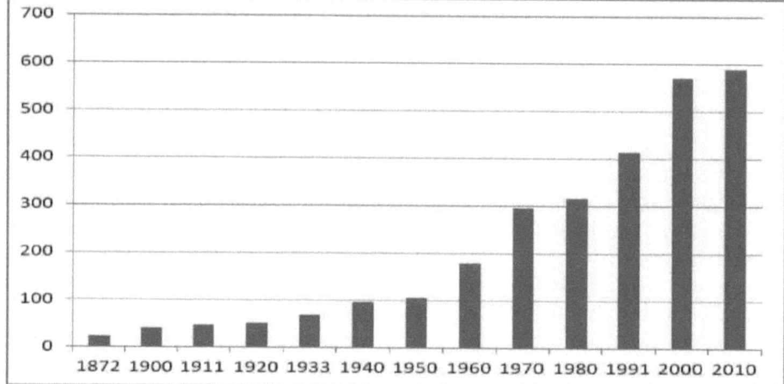

GRAPH 02 - CHANGES IN THE NUMBER OF BORDER MUNICIPALITIES BETWEEN 1872 AND 2010
SOURCE: IBGE. Own elaboration

However, although today the border strip represents 27% of the national territory, the ratio between the number of border and non-border municipalities indicates that all border municipalities are relatively small compared to the number of municipalities in Brazil. The relationship between the two variables shows that from the 1940s onwards, the percentage difference became more significant. One of the reasons for the increase in the number of border municipalities was the expansion of the border strip (Graph 03).

In any case, it can be analyzed that the rate of dismemberment in the HF is similar to the process that took place in the rest of the country. The periods of greatest creation of municipalities were between 1960 and
1970 and between 1991 and 2000, with the emergence of 281 new municipalities, representing 47% of the current total. The number of cities doubled between 1900 and 1911 and tripled between 1950 and 1980, so in just 80 years the number of municipalities increased more than fivefold. As Pumain (1997) points out, regardless of the country's level of development, the city system always tends to grow, both in terms of urban population and the number of municipalities.

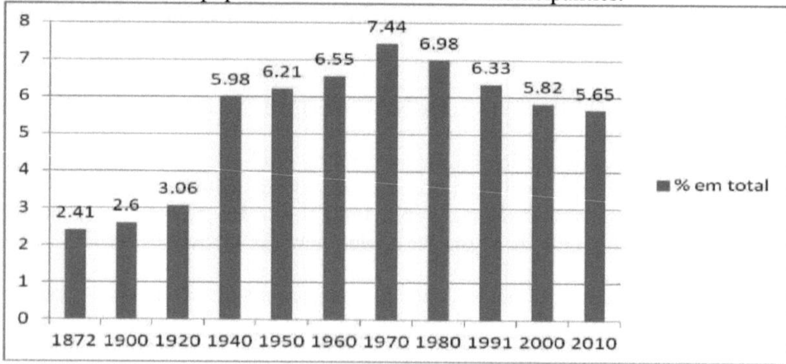

GRAPH 3 - RELATIVE PERCENTAGE OF BORDER AND NON-BORDER MUNICIPALITIES
SOURCE: IBGE. Own elaboration

The pace of city break-ups on the border strip is very similar to that of the rest of the country. The period of greatest dismemberment occurred in the 1960s, when 117 municipalities were created, and between 1991 and 2000, when 156 municipalities were created. On a national scale, the period of greatest division was between 1950 and 1970, and it was also intense between 1960 and 1970, when the national network increased by 46%.

According to the IBGE, the construction of the urban fabric in Brazil was influenced by different political moments, according to the political interests of the federal and state governments, and presented different rhythms in each decade. The main cause of the emancipation of municipalities between 1950 and 1960 was the application of the Municipal Participation Fund (FPM), introduced by the 1946 Constitution, with equal quotas for all municipalities. As a result, some state governments encouraged the creation of municipalities in order to obtain more federal funding (IBGE, 2010).

With regard to the type of municipality created, there are three specific situations: dismemberment, annexation and litigation. There were approximately 515 dismemberments on the border strip, of which 54 were in the Northern Arc, 85 in the Central Arc and 376 in the Southern Arc. The only annexations took place in Clevelândia (PR) and Sâo Martinho (RS). The only case of litigation was in the municipality of Amapá, which was inserted into Brazilian territory after the litigation between Brazil and France was finalized.

Comparatively, between 1872 and 2010, in territorial terms, it can be said that the border territory underwent a profound process of transformation due to the following factors:

a) processes of border disputes, with the inclusion or loss of territories[24] ; as a consequence, new municipalities were included or excluded;

b) intense process of dismemberment and creation of new municipalities, motivated by various factors, in different decades. The settlement of the FF is rectilinear and the municipal division is more dynamic, so there have been some variations in the total number of municipalities;

c) formation of border municipalities derived from a few "temtórios-maes", giving rise to new municipalities through the process of dismemberment and multiplying the number of border towns, mainly in the states of Paraná and Santa Catarina;

d) changing the width of the border strip (66km - 100km - 150km) included more border municipalities, in accordance with current legislation.

4.2 Border dispute processes

The result of the processes of territorial litigation and the acquisition of territories has been the inclusion of new territories and, consequently, new municipalities have emerged in these areas. For example, the Palmas Question included 30,612 km^2 . In 1872, the municipality of Guarapuava (PR), until then, was the only municipality in Paraná that touched the border. After the conflict was resolved, the territory increased and Sao Joao do Capanema and Bella Vista were emancipated from Palmas.

In the Amapá Question, 31,650 km^2 of surface area were annexed, which gave rise to the municipality of Macapá, elevated to the category of town by State Law No. 798, of October 22, 1901 and installed on April 30, 1902 (IBGE-cidades, 2014). At the time, the surface area of Macapá was 96,316,605 km.[27]

In the case of Acre, between 1900 and 1911 the main territorial change was the official inclusion of the territory of Acre and its transformation into a federal territory. The Federal Territory of Acre had only 3 municipalities, all of which bordered it: Alto Juruá, Alto Purús and Alto Acre, but none had urban headquarters on the border strip.

4.3 Municipal dismemberment processes on the borders

In the southern region, the states of Paraná, Santa Catarina and Rio Grande do Sul have had a large number of border municipalities since 1872, compared to the other states.

---

[27][24] The processes of territorial disputes and the acquisition of territories are described in more detail in Chapter 3 of this thesis. The main ones were the Palmas Question, the Amapá Question and the acquisition of Acre.

Rio Grande do Sul was the state in which the process of dismemberment was most intense among all the border states. In 1872, 43% of all border municipalities were from Rio Grande do Sul, while in 2010 they accounted for 33.5% of the total. In absolute terms, Rio Grande do Sul has the largest number of municipalities of all the border states, followed by Paraná and Santa Catarina (Graph 04). Santa Catarina's border municipalities were only included in the IBGE's territorial division in the 1920 census, when the territorial dispute between Paraná and Santa Catarina was resolved.

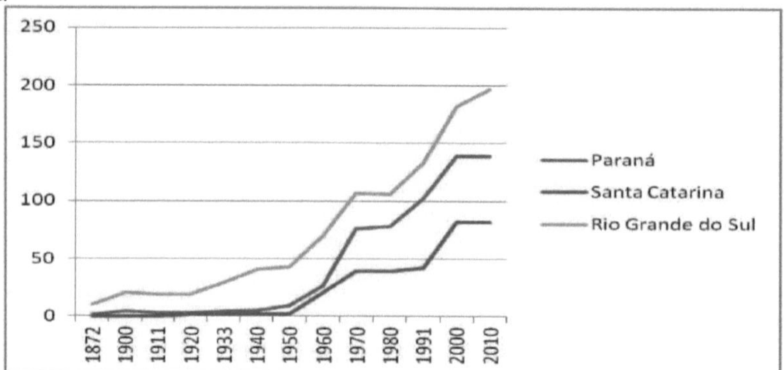

GRAPH 04 - TEMPORAL EVOLUTION OF THE NUMBER OF BORDER MUNICIPALITIES IN THE STATES OF SANTA CATARINA, PARANÁ AND RIO GRANDE DO SUL
SOURCE: IBGE. Own elaboration

In 1872, the only border municipality in the state of Paraná was Guarapuava, which had no urban center on the border strip. This means that there was officially no urban center on the 66km border strip. Although Guarapuava is not currently considered a border municipality, until 1950 its territory touched the border strip. Guarapuava had a surface area of 121,142,287 km^2 , comparatively, only this municipality was larger than Portugal.

The Palmas region was in the process of an international dispute with Argentina. Matos (1990) states that military towns were set up with the aim of defending and populating the territory of Palmas, in order to mark the military presence in the area. The military colonies created were Foz do Iguapu, Chopim, Chapecó and Alto Uruguai.

In 1872, Rio Grande do Sul already had the largest number of border towns, including urban centers on the border strip. The municipalities that are now considered to be twin cities, such as Itaqui, Jaguarao, Livramento, Sao Borja and Uruguaiana, were already present in this census. The municipality of Bagé, although not a twin city, already had its municipal seat on the border.

Rio Grande, Cruz Alta, Alegrete and Piratini were cities that had international boundaries, but their municipal headquarters were located outside the 66 km width of the border strip. Of the 23 border towns, 9 had their municipal seat within 60 km and most were in Rio Grande do Sul (6 in total).

In 1990, new municipalities were created on the Paraná border strip, among them: Bella Vista de Palmas (now Clevelandia), Palma and Sao Joao do Capanema. Palma with a surface area of 31,768,361 km^2 and Bella Vista de Palmas with 8,192,809 km2, both municipalities being dismembered from Guarapuava and Sao Joao do Capanema dismembered from Bela Vista.

In Rio Grande do Sul, the number of border municipalities doubled between 1872 and 1900, with Arroio Grande, Dom Pedrito, Herval, Palmeira, Cacimbinhas, Quarahy, Rosário, Santa Victoria do Palmar, Santo Ángelo, Sao Luiz Gonzaga and Sao Martinho emerging. The latter was the smallest municipality on the border with Argentina, with just 172,667 km2, surrounded by neighboring Santo Ángelo, with 10,077,927 km2, and Palmeira, with 10,164,839 km2. Quarahy, which is now considered a twin city, was the only one in the state of Rio Grande do Sul to be emancipated after 1872.

In 1911, the municipality of Sao Joâo do Capanema was annexed to Clevelândia, and there was also the annexation of Sao Martinho, in Rio Grande do Sul, the only cases of indexation that occurred on the border strip from 1872 to 2010. Thus, the number of border municipalities between 1900 and 1911 in Paraná and Rio Grande do Sul was reduced.

In the 1920 census, the territory of Santa Catarina took on its current shape after the end of the dispute. The only border municipality was Chapecó, with an area of 13,938,795 km^2 , which was the "mother territory" of all the border towns that emerged later. In Rio Grande do Sul and Paraná there was no change. And in Paraná, Foz do Iguaçu was emancipated from Guarapuava with an area of 19,539,539 km2. According to IBGE-Cidades (2014), Foz do Iguaçu had been founded as a military colony in 1889, marking the beginning of the effective occupation of the place by Brazilians, and which would become the municipality of Foz do Iguaçu. In 1910, the military colony became "Vila Iguassu", a district of the municipality of Guarapuava. Two years later, the Minister of War emancipated the Colony, making it a civilian settlement entrusted to the care of the government of Paraná, which then created the Village's State Collectorate. It was elevated to the status of a city, with the name of Foz do Iguaçu, by State Law No. 1,658 of March 3, 1917. In the administrative division for 1933, the municipality is made up of the main district.

Between the 1930 and 1940 censuses, the border strip was extended. As a result, some existing municipalities were included. In the first expansion there were 10 municipalities and in the second there were 12 municipalities. In addition, between 1940 and 1950, two new municipalities were created in Rio Grande do Sul: Santa Rosa, separated from Santo Ângelo, and Iraí, separated from Palmeira, both of which had their urban headquarters on the strip.

In 1943, the Federal Territory of Iguassú was created, influenced by Backheuser's border policy. After the 1930 Revolution, Backheuser's ideas produced a policy of strengthening the border regions, based on the concept that the border is the epidermis of the state organism, capturing foreign influences and pressures and, as such, should be subordinated to the central power and not to the regional authorities, who are less sensitive to these problems. In short, border policy should not be regional, but federal (MATTOS, 1980). In this way, the municipalities of Foz do Iguassú, Clevelândia, Iguassú, Mangueirinha and Chapecó formed the Federal Territory of Iguassú, subordinated directly to the central government.

In 1950, the state of Paraná was further broken up. Pitanga and Laranjeiras do Sul emerged from Guarapuava. Guarapuava continued to touch the border. Pitanga was then broken up to form Campo Mourao, with an area of 25,934,843 km^2 . As a result, Pitanga and Campo Mourao no longer share a border.

In the northernmost region of the state, Mandaguari becomes a borderland, coming from Apucarana, which came from Londrina, and Londrina is no longer on the border strip. Mangueirinha also emerged as a border territory, originating from Palmas, with 4,296,099 km2. From then on, Palmas and Guarapuava touch the minimum on the border. Foz do Iguaçu, Campo Mourao, Mangueirinha, and the western Mandaguari are the "mother cities" from 1950 onwards, replacing Palmas and Guarapuava.

From 1960 onwards, the process of municipal creation intensified in the southern region. In Paraná there were 17 dismemberments, originating in Clevelândia, Foz do Iguaçu, Campo Mourao, Mangueirinha and Paranavaí. According to IBGE Cidades, Capanema became a municipality again - it had already been a municipality in 1900 - and was dismembered from Clevelândia (formerly Bella Vista de Palmas).

Between 1950 and 1960, the process of territorial division began on the border of Santa Catarina, which until the previous decade had only had two municipalities. From then on, it multiplied and 16 municipalities were created, originating from Chapecó, Concórdia and Dionísio Cerqueira (it was elevated to the category of municipality with the name of Dionísio Cerqueira, by state law no. 133, of 30-12-1953, dismembered from Chapecó - IBGE cities, 2014).

In Rio Grande do Sul, the 26 dismemberments originated from Erechim (previously 4,277,508 km^2 , now 1,706,170 km^2), Palmeira (previously 5,282,788 km2 , now 2,590.317 km2) , Três Passos (previously 4,219,183 km2, now 680,789 km2) , Santa Rosa (489,086 km2, previously 4,028,317 km2). Erechim, Três Passos and Palmeira are located in the northwest of Rio Grande do Sul.

Between 1970 and 2010, Paraná created 63 border municipalities, Santa Catarina, 62 border municipalities and Rio Grande do Sul 128 border municipalities. The southern region leads the *ranking in terms of the* number of border municipalities. The following chapter will discuss the demographic growth of this region and the demographic change in these cities.

In the central-western region, the state of Mato Grosso was the second largest border state in 1872 and until then there was no state of Rondônia or Mato Grosso do Sul. The evolution of the creation of border municipalities in the states of Mato Grosso, Mato Grosso do Sul and Rondônia can be seen in Graph 05.

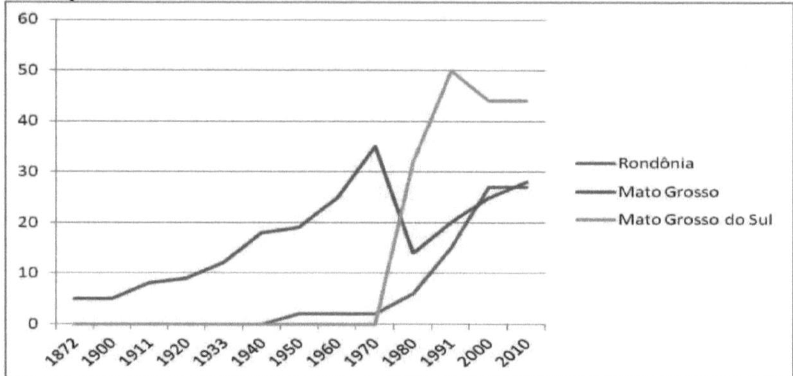

GRAPH 05 - EVOLUTION OF THE CREATION OF BORDER MUNICIPALITIES IN THE STATES OF MATO GROSSO, MATO GROSSO DO SUL AND RONDÔNIA
SOURCE: IBGE. Own elaboration

The cities with urban headquarters in the belt were Corumbá and Matto
The latter was later dismembered and part of the territory became the state of Rondônia. In 1872, Mato Grosso had a total of 5 border municipalities, 2 of which were located on the border and 3 of which were not, namely Miranda, Poconé and Vila Maria.

It's interesting to note that the state of Matto Grosso was difficult to reach. Transportation was done by horse and it took days to get there. Another means of transportation was by river via the Paraguay and Paraná rivers[28] . In this way, the population was isolated from the rest of the territory. According to Matos (1990), some military colonies were set up in Nioac (1854), Rio Brilhante (1854) and Dourados (1856) with the aim of surveillance, military protection and also settlement.

In 1911, two municipalities emerged in the state of Mato Grosso: Porto Murtinho (dismembered from Corumbá) and Bella Vista (dismembered from Nioac, a former military colony), being municipalities based on the border strip close to Paraguay. Nioac was dismembered from Miranda and is no longer a border town.

In 1920, Mato Grosso created the municipality of Ponta Pora (originally from Bella Vista, which had an area of 56,552,082 km^2), with an area of 46,500,691 km2, meaning that Bella Vista had a smaller territory of 10,051,391 km2. Today, Ponta Pora is considered a twin city.

In 1933, the state of Mato Grosso was further dismembered. Guajará Mirim, which was dismembered from Santo Antonio do Rio Madeira, had a vertically shaped territory, a large part of which (approximately 633 km) was located on the international border with Bolivia. The surface area of this new municipality was 89,584,538 km2. The other case was Maracajá, with territory originating from Nioac. In 1940, in Mato Grosso, there were new dismemberments of 6 cities, concentrated on the border with Paraguay.

In 1943, the Federal Territory of Ponta Pora was created, divided into 7 municipalities: Porto Murtinho (17,758,857 km^2 of surface area), Bela Vista (10,051,391 km2), Ponta Pora (25,932,939

---

[28] This has already been covered in the Palmas Question in Chapter 3.

km2), Dourados (20,567,752 km2), Miranda (12,930,139 km2), Nioaque (5,137,842 km2) and Maracajú (5,298,028 km2), the last being the capital of the federal territory, which was abolished in 1946. The Territory of Guaporé was made up of the following municipalities: Porto Velho (23,941,399 km2 of surface area), Alta Madeira (280,887,183 km2) and Guajará Mirim (89,584,538 km2), with Porto Velho as the capital. In 1956, the name of the Federal Territory of Guaporé was changed to the Federal Territory of Rondonia, becoming a unit of the Federal Government with the 1988 Federal Constitution. The Federal Territory of Rio Branco became the Federal Territory of Roraima and a unit of the Federation, together with Rondonia.

As long as the Federal Territory of Rondonia existed, there were no dismemberments on the border strip. After 1980, the process of municipal creation began in Rondonia, with a 22% increase in the number of border municipalities between 1980 and 2010.

The state of Mato Grosso do Sul was officially created in 1979. Its cities previously belonged to the state of Mato Grosso. According to the 1970 census, there were 34 border towns in Mato Grosso. In the 1980 census, the number dropped to 14 due to the division of the state. On the other hand, Mato Grosso do Sul, then recently created, incorporated 34 border municipalities. Today, 53% of the municipalities in the Central Arc belong to Mato Grosso do Sul.

Before the division, the state of Mato Grosso had a surface area of 1,262,143,326 km^2 , with the creation of Mato Grosso it now has 904,840,502 km^2 ' losing 41% of its border towns. Today, Mato Grosso has only 4.76% of all Brazilian border towns.

In the northern region, the process of creating border municipalities had some particularities. The first is that, in general, the land area in the region is much larger than in the south, so when the border strip was extended (from 66km to 100km and then to 150km) the number of border municipalities did not increase significantly. The second is that more federal territories were created in the north than in the center-west and south, which influenced the number of municipalities created, because while they belonged to the federal territory, the number of border municipalities was small. Graph 06 shows the evolution of border municipalities in the border states of the northern region.

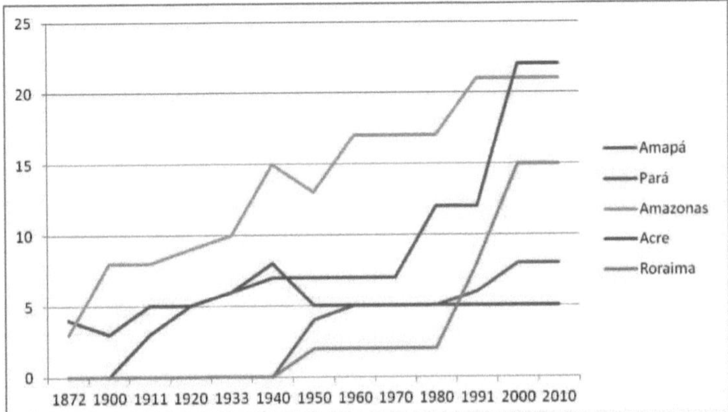

GRAPH 06 - EVOLUTION OF BORDER MUNICIPALITIES IN THE STATES OF AMAPÁ, PARÁ, AMAZONAS, ACRE, RORAIMA
SOURCE: IBGE. Own elaboration

In 1872, Acre did not belong to Brazil. It only became a federal territory on November 7, 1903, when the Treaty of Petrópolis was signed to purchase 191,000 km2 of Bolivian territory. The cities of the Amazon emerged in the 19th century at strategic points of navigation on the main rivers. The cities served as support for the activities that exploited the region's economic potential and also as defense points. The current territory of the state of Roraima belonged to the state of Amazonas, specifically the municipality of Manáos.

The municipalities that had contact with the international borders were: Manáos, with an area of 883,129,764 km2, bordering Venezuela and Guyana; Tefé, with 524,517,291 km2, bordering Colombia and Peru; and Barcelos, with 272,429,121 km2, bordering Colombia and Venezuela. All of them had their urban centers well away from the border and only part of the municipality's surface was on the international boundary. At that time, there were no urban centers on the border strip in the northern region and the border municipalities had a very large territorial area.

The municipalities of Gurupá, Óbidos[29] , Macapá and Mazagao belonged to the province of Pará and bordered present-day Guyana, French Guiana and Suriname. Until then, the state of Amapá did not exist. Its territory belonged to Pará and no municipality had its urban headquarters on the border.

There were no major changes between 1872 and 1900. The territory of Acre was still under negotiation with Bolivia and the states of Amazonas and Pará remained the only border states in the northern region.

In 1911, in the state of Amazonas, there were new dismemberments along the border, and the number of municipalities more than doubled in 28 years. Rio Branco, which later became the capital of Acre, at this time belonged to the state of Amazonas, with an area of 226,046,342 km^2 . Of the 3 original border municipalities (from 1872), only 2 remained border municipalities: Barcellos, with an area of 95,819,356 km2, and Tefé, with 244,960,766 km2. Manáos gave rise to other municipalities and its territory is no longer in the border strip.

The only urban center in the FF of Amazonas and the northern region in 1991 was the municipality of Floriano Peixoto, now Boca do Acre, which, according to the history of the municipality published by the IBGE (IBGE Cidade, 2014), was born under the influence of the rubber cycle and was previously inhabited by the Capanas and Aripuanas Indians. After the "acquisition of Acre", Boca do Acre is no longer a border town. Floriano Peixoto was close to the international border with Bolivia and Peru and its territory was dismembered from Labrea.

In Pará, the only new town was Almeirim, which was dismembered from Gurupá (no longer a border town), i.e. in quantitative terms there was a replacement. In Mato Grosso there was a similar case: Nioac was dismembered from Miranda, no longer a border town.

Between 1900 and 1911, the main territorial change was the official inclusion of the territory of Acre in Brazilian territory. In this way
The Federal Territory[30] of Acre was born with 3 municipalities, all on the border: Alto Juruá, Alto Purús and Alto Acre, but none of them had an urban seat on the border, the nearest being Alto Acre. Floriano Peixoto, which was part of the border strip, ceased to be a border town when Acre was included. After the Acre process was completed, Brazil began a demarcation process with Peru.[31]

In the state of Amazonas, the two main changes were the renaming of Rio Branco to Boa Vista do Rio Branco and the emergence of Benjamin Constant, dismembered from Sao Paulo de Olivenpa. In December 1900, the Amapá dispute was finalized, officially incorporating 31,650km^2 into Brazilian territory.

Between 1911 and 1920, all the municipalities in Acre changed their names and Villa Seabra (originally from Alto Jaruá) and Xapuri (from Alto Acre) were emancipated. In Amazonas, Sao Fellipe (from Tefé) and Porto Velho (from Humaitá) were also emancipated. Humaitá, which used to be a border town, is no longer so.

In 1993, the municipalities of Feijó and Sena Madureira were created in Acre, and again 50% of the municipalities were renamed. All of Acre's municipalities are border municipalities and only two are located on the border. In Amazonas there is a curious case: Floriano Peixoto once again touches the border strip because the width of the strip has increased. After the purchase of Acre, the municipality had ceased to be a border municipality because its territory had increased and the measurement between the international boundary and the width of the previous border strip did not

---

[29] Óbidos was set up as a military colony in 1854 and established by the Empire's military surveillance border policy (MATOS, 1990).

[30] Federal territory was a kind of legal-normative prosthesis discussed in Porto's work (2003). Details can be found in Chapter 3 on the emergence of federal territories.

[31] See chapter 3 on the Acre Question.

include Floriano Peixoto.

In Pará, the municipality of Montenegro, which is in the far north of Brazil, was renamed Amapá, with an area of 63,178,262 km^2 . Amapá had been the capital of the Federal Territory of Amapá, created by Law No. 798 of October 22, 1901. An important municipality, due to its strategic geographical position, and also because this territory was contested by France and the final outcome of the dispute was favorable to Brazil. In the state of Pará, the municipality of Oriximiná was created, separated from Óbidos, whose territory lost 80.15% of its size to the new municipality.

In 1940, new municipalities emerged on the border strip. Brasilia, dismembered from Xapury, in Acre, and Mazagao do Amapá, from Macapá, from which it inherited 58% of the territory.

Porto Velho belonged to the state of Amazonas and had a surface area of 23,941,399 km2, Guajará Mirim belonged to Mato Grosso and had a surface area of 89,584,538 km2, of which approximately 639,810.10 km was the municipality's border with the international border. After the creation of the Federal Territory of Guaporé, Porto Velho now has an area of 150,792,089 km2, i.e. an increase in its territory of 68%, while Guajará Mirim remained the same as before (89,584,538 km2). The municipal territory of Porto Velho originated from Aripuana, which reduced its territory by 54%. Before it was created, it was called Alto Madeira.

In Acre and Amazonas there were no new dismemberments, the only territorial change between the 1933 and 1940 censuses was that Boa Vista, with 95,381,687 km2, no longer belonged to the state of Amazonas. In Pará, with the creation of the Federal Territory of Amapá, the municipalities that previously belonged to Pará (Macapá, Mazagao and Amapá) became part of the federal territory by Decree-Law No. 6,550, of
May 31, 1944. Amapá lost its capital to the municipality of Macapá. Amapá was also dismembered, giving rise to the municipality of Oiapoque, with an area of 22,646,098 km^2 .

Between 1950 and 1960, there were new divisions in Amazonas and some towns no longer touched the border strip, including Carauari, Eirunepé, Parintins and Tefé: Carauari, Eirunepé, Parintins, Tefé. In Pará, the municipality of Amapá was broken up again, and Calçoene was created, with an area of 14,276,542 km2. Between 1960 and 1970, there were no border municipalities between the states of Amazonas, Pará, Acre, Amapá and Roraima.

The dismemberment of municipalities had different rhythms in each state. Until 1980, Roraima had only 2 municipalities; then 13 more border municipalities came into being and since then there have been no more break-ups. It's interesting to remember that from 1943 until it became a unit of the Federation with the 1988 Constitution, Roraima was a Federal Territory, made up of the municipalities of Boa Vista and Catrimani. Roraima is the state with the fewest municipalities in Brazil.

While it existed, the Federal Territory of Amapá created only one municipality on the border strip. After the 1988 Constitution there were more municipal dismemberments, mainly between 1991 and 2000. In the state of Acre there was a similar case, after being included as a unit of the Federation the process of dismemberment of Acre was more intense between 1970 and 1980 and between 1991 and 2000.

In Amazonas, the highest rate of creation of municipalities was from 1933 to 1940, but the state lost territory when the Federal Territory of Rio Branco was created. Amazonas is the state in the Northern Arc with the second highest number of municipalities and also the highest number of towns located in a border region.

The border municipalities in the state of Pará were more numerous in the 1940s, and the state also lost territory after the creation of the Amapá Territory. In Pará, despite its large territorial area, only 5 municipalities are border municipalities and 2 municipalities have urban headquarters on the border strip.

4.4 Distribution of border municipalities by arc

Currently 71 border municipalities are located in the Northern Arc, 99 in the Central Arc and 418 in the Southern Arc. The distribution of border municipalities by border arc between 1900 and 2010 can be seen in Graph 07.

The Southern Arc is made up of the states of Rio Grande do Sul, Paraná and Santa Catarina, and has had the largest number of border municipalities since 1900. Rio Grande do Sul accounted for more than half of the Arco Sul's border municipalities until the 1960s. Currently, the highest concentration of border municipalities remains in the Southern Arc, with 33% in Rio Grande do Sul, 23% in Paraná and 13% in Santa Catarina.

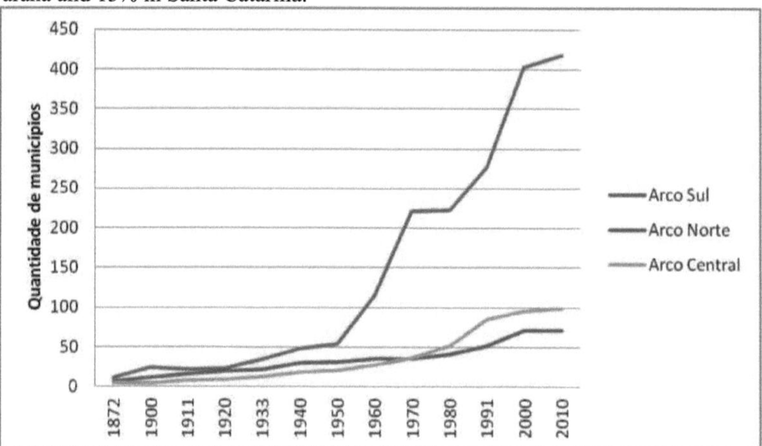

GRAPH 07 - DISTRIBUTION OF BORDER MUNICIPALITIES BY BORDER ARC BETWEEN 1872 AND 2010

SOURCE: IBGE. Own elaboration

On a national scale, Rio Grande do Sul is the third state with the largest number of municipalities, Paraná ranks fifth and Santa Catarina sixth, meaning that in these states the large number of municipalities is not restricted to the border strip.

The period of greatest emancipation in the Central Arc, made up of the states of Mato Grosso do Sul, Mato Grosso and Rondônia, occurred between 1980 and 2000. It is the second arc to concentrate the largest number of border municipalities.

In Arco Norte, the largest number of municipalities was created between 1991 and 2000, after the 1988 Federal Constitution, which transformed the federal territories into federal units. The state of Rondônia only officially appeared in the territorial division of 1950, until then this region belonged to Amazonas and Mato Grosso. In Rondônia, 52% of the municipalities are border towns.

In Arco Norte, the political and administrative changes were more profound because several federal territories inspired by US law were created (SOUZA *et al*, 2014), such as the current states of Acre, Roraima and Amapá. In addition, the region has a low population density and the border municipalities have a large surface area. Another particularity is that some states in the Northern Arc are completely on the border, such as Roraima and Acre, which is not the case in other arcs.

4.5 Formation of municipalities derived from large territories - mothers

In the southern region, the area of border municipalities was already smaller compared to other provinces. The largest municipalities in 1872 were Matto Grosso (430,782,098 km2) and Manáos (883,129,764 km2). By way of comparison, in Rio Grande do Sul, Cruz Alta (33,850,443 km2) was the largest border municipality, 12 times smaller than Matto Grosso and 26 times smaller than Manáos.

The great "mother city" of the border towns in the states of Paraná and Santa Catarina was the municipality of Guarapuava (PR), until the end of the dispute between Paraná and Santa Catarina, and there was no urban seat on the border strip. Although today Guarapuava is not considered a border municipality, until 1960 its municipal territory touched the border strip. In 1872, the municipality of Guarapuava had a surface area of 121,142,287 km^2 , comparatively larger than Portugal.

4.6 Inclusion of new municipalities due to changes in the width of the border strip

The 1930 territorial division saw the first extension of Brazil's border strip, from 66km to

100km, i.e. the inclusion of 34km of width. As a result, new municipalities were added to the existing border strip. Classified in this case are 1 municipality in Pará, 1 in Paraná and 9 in Rio Grande do Sul. In total, this represents a 33% increase in the number of border municipalities. In addition, there was a new split in Rio Grande do Sul with the creation of Pinheiro Machado, split from Piratini.

The 1930 territorial division saw another extension of Brazil's border strip, from 100 km to 150 km, i.e. 50 km wide. Classified in this case are 5 municipalities in Amazonas, 1 in Pará, 12 in Rio Grande do Sul and 4 in Mato Grosso. This represents a 41% increase in the number of border municipalities. In addition, there were new break-ups in Santa Catarina (1), Paraná (1), Acre (1) and Mato Grosso (2).

4.7 The presence of twin cities in Brazil's first census

The first Brazilian census was carried out in August 1872 by the *General Directorate of Statistics* (DGE) during the imperial period, which lasted only 67 years. The transitional period between the last years of the monarchy and the republic was marked by major socio-economic changes such as the expansion of coffee growing, European migration and the end of slavery. The 1872 general census took place precisely in this political context.

It is the only census to provide information on slavery and the lack of quality of previous attempts at counting is explained by the population's fear of the tax authorities and military conscription. The 1872 census was carried out in 20 provinces, distributed in 641 cities and 1 "neutral municipality" (now the city of Rio de Janeiro) (CEDEPLAR, 2012).

The parishes were responsible for counting the population distributed over the geographical area. There were a total of 1,440 parishes and each one had its own committee. The census collected the number of free men and slaves distributed according to gender, race, marital status, religion, physical defects, nationality and level of education.

At the time, Brazil's total population was around 10 million, 15% of whom were still slaves. Brazil's population was concentrated on the coast and in the southeast. The most populous provinces were Minas Gerais, Bahia and Pernambuco.

The "border provinces"[32] , in this case, include Rio Grande do Sul, Paraná, Matto Grosso, Amazonas and Grao Pará, which had very different territorial dimensions to the current ones[33] and were also not the most populous in Brazil.

With regard to slavery in the border provinces, 32.47% of the workforce was made up of slaves, while on a national scale this represented 24.87% of all slaves. The province with the largest slave population was Pará, followed by Rio Grande do Sul and Santa Catarina. Table 06 shows the slave and free populations of the border provinces in 1872.

TABLE 06 - SLAVE AND FREE POPULATION OF THE BORDER PROVINCES IN 1872

Provinces	Number of municipalities	Pop. Free	Pop. Slave	Total
Rio Grande do Sul	28	367,022	67,791	434,813

---

[32] These are considered to be the provinces that had international boundaries with other countries.
[33] See Table 5.1 - Surface area of border states in 1872 and 2010.

Amazonas	7	56,631	979	348,009
Santa Catarina	11	144,818	14,984	159,802
Paraná	16	116,162	10,560	126,722
Mato Grosso	9	53,750	6,667	60,417
Pará	32	274,779	274,779	27,458
Total selected provinces	103	1,013,162	375,760	1,157,221
Total Brazil	641	8,419,672	1,510,806	9,930,478

Source: CEDEPLAR, 2014

This census indicates that there were 6 border municipalities that are now classified as twin cities, located in the provinces of Rio Grande do Sul and Matto Grosso. In the northern region of Brazil at that time, there were no municipal seats located close to the international border, but rather a few "villas"[34] .

The border towns of Livramento (Sant'Ana do Livramento today), Itaquy (Itaqui), Jaguarao, Sao Borja and Uruguayana (Uruguaiana) belonged to the province of Rio Grande do Sul, which in 1872 had 28 towns in its territory. After a few years, it became a federal state with the promulgation of the first Federal Constitution of the Republic in 1891.

Corumbá belonged to the province of Mato Grosso, which at the time was made up of nine cities and later became a federal state. On November 11, 1977, the territory of Mato Grosso was divided into two parts, giving rise to the state of Mato Grosso do Sul, in which the municipality of Corumbá is currently located.

The population of the border municipalities of Rio Grande do Sul was 54,591 (12.55% of the province's total population) and Corumbá represented 5.56% of the total population of Matto Grosso. The order of greatest importance in relation to the population of the selected municipalities is Sao Borja, Jaguarao, Sant'Ana do Livramento, Itaqui, Uruguaiana and Corumbá.

In the border towns, slavery still accounted for 14% of the total population. There were a total of 8,537 slaves, of whom 4.5% were men and 3.5% women. The municipality with the largest slave population was Jaguarao, with 3,248 slaves, representing 23% of the total population, followed by Sant'Ana do Livramento (19%) and Sao Borja (7.69%). The distribution of the free and slave populations of the selected cities can be seen in Table 07. The selected cities in Rio Grande do Sul accounted for 12% of the total number of slaves in the province of Rio Grande do Sul. Corumbá was home to 4% of the total number of slaves in the province of Matto Grosso.

TABLE 07 - DISTRIBUTION OF THE POPULATION IN THE SELECTED MUNICIPALITIES

Municipalities	Pop. Free	Pop. Slave	Pop. Total	Born in Brazil	Estra	Pop. Total
Corumbá	3.086	275	3361	2697	664	3361
Itaqui	7.697	864	8561	7795	766	8561
Jaguarâo	10.514	3248	13762	11454	2308	13762

---

[34] For more details, see chapter 5.

Sao Borja	**13.686**	**1141**	**14827**	**14076**	**751**	**14827**
Santana do Livramento	**8.063**	**2012**	**10075**	**9464**	**611**	**10075**
Uruguaiana	**6.369**	**997**	**7366**	**6511**	**855**	**7366**
Total	**49.415**	**8537**	**57952**	**51997**	**5955**	**57952**

Source: CEDEPLAR, 2014

On the other hand, the free population in the selected cities was 49,415 inhabitants (85% of the total population), made up of 52.66% men and 47.33% women. The largest population of free men was in Sao Borja (located on the border with Argentina), followed by Jaguarâo and Sant'Ana do Livramento (located on the border with Uruguay).

Another factor that affected demographic dynamics was the arrival of European migrants, attracted by the offer of jobs on the coffee plantations. The nationalities of the immigrants in the border provinces, in order of importance, were: German, Portuguese, African, Oriental[35] , Paraguayan, Italian, Argentinian, French, Spanish and English.

The border province that received the most immigrants was Rio Grande do Sul, mainly of the German, Portuguese and Oriental nationalities. The selected city that received the largest number of foreigners was Jaguarao, followed by Uruguaiana and Itaqui.

The highest immigration rate was in Corumbá with 19.75%, followed by Jaguarao, 16.77%, and Uruguaiana, 11.60%. In Sant'Ana do Livramento, Uruguaiana and Jaguarao, the most common nationality was Oriental. In Sao Borja and Corumbá it was Paraguayan, in Itaqui it was Argentinian.

Since the proclamation of the Republic of Brazil in 1889, the census has become an important tool for electoral issues inherent in the new reality, as well as being useful for visualizing the demographic distribution of Brazilian territory.

4.8 Spatial distribution of the border population (1900-2010)

Population growth on the border between 1900 and 2010 was on the rise. As Pumain (1997) points out, regardless of the country's level of development, cities always have a tendency to grow in terms of urban population.

According to the latest census in 2010, the municipalities along the border have approximately

---

[35] Although oriental is not a nationality, it is how it was classified by the database.

11 million inhabitants, which represents only 5.65% of the total Brazilian population distributed over 27% of the national territory. Thus, the border municipalities as a whole have a disproportionate characteristic because, compared to the rest of the country, they have a low population concentration, with the population spread over a vast territory. On the other hand, in 110 years (between 1900 and 2010) the border population increased 21 times. Graph 08 shows the population growth curve of the border population between 1900 and 2010.

The second characteristic is that the concentration of the border population is not homogeneous; it is more concentrated in the Southern Arc compared to the others. Consequently, the Southern Arc is also home to the largest number of municipalities, as discussed in this chapter.

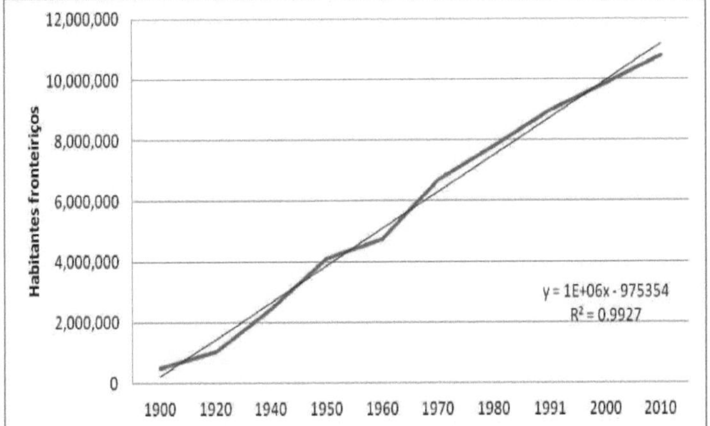

GRAPH 08 - BORDER POPULATION GROWTH CURVE BETWEEN 1900 AND 2010
SOURCE: IBGE. Own elaboration

Generally speaking, in the border municipalities the period of lowest growth, in absolute terms, was between 1900 and 1920 and the period of highest population growth was between 1960 and 1970. In the Southern Arc, growth accelerated from 1920 onwards, associated with the occupation policy known as the "march to the west", during the Getúlio Vargas government. In the Central and Northern Arcs, there was an increase in population from 1970 onwards. The population of the Central Arc was smaller than that of the Northern Arc until the 1960s, then, with the expansion of agriculture in the states of Mato Grosso and Mato Grosso do Sul, it became larger than that of the Northern Arc (Graph 09).

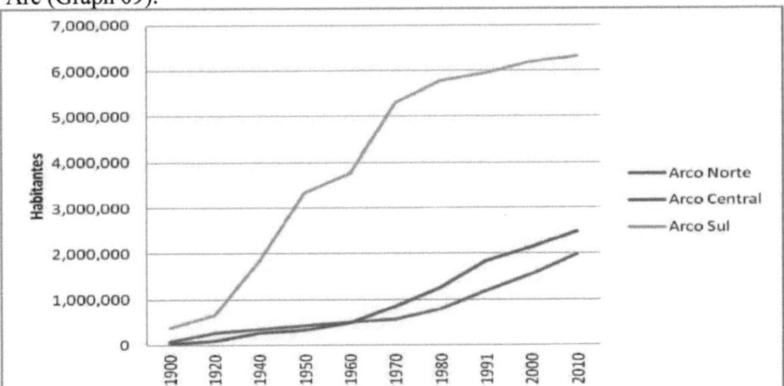

GRAPH 09 - BORDER POPULATION GROWTH BETWEEN 1900 AND 2010 DISTRIBUTED IN BORDER ARCS

SOURCE: IBGE. Own elaboration

The highest average growth rate in Arco Norte was between 1970 and 1980, with rates of 1,025%. This growth may be related to government incentives to promote economic development in the northern region.

In Arco Central and Arco Sul, the highest growth rate was between 1940 and 1950, with rates of 1,031% and 1.01% respectively. Since 1950, the growth rate in the Northern Arc has decreased with each census. The average population growth rates[36] by border arch and by period between censuses can be seen in Graph 10.

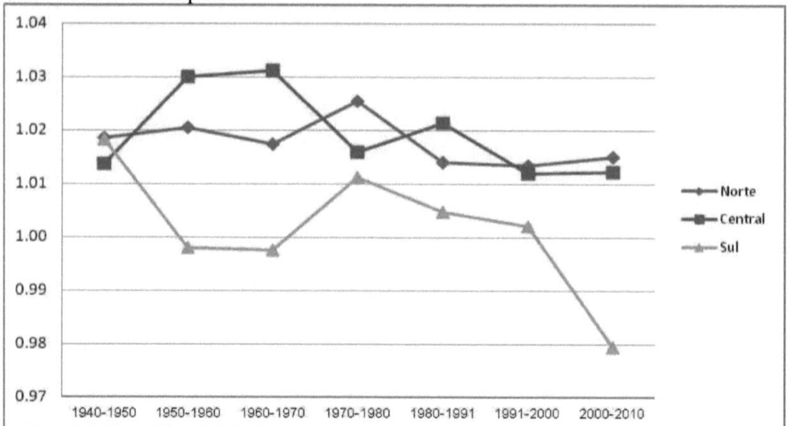

GRAPH 10 - AVERAGE POPULATION GROWTH RATES BY BORDER ARC BETWEEN 1940 AND 2010

SOURCE: IBGE. Own elaboration

In comparison, the average annual growth rate of the
between the 2000 and 2010 censuses is 1.17%. The current average growth rate for border municipalities is 1%, with 0.97% in the Southern Arc and 1.01% in the Northern and Central Arcs, i.e. lower than the national average.

The population growth of the border municipalities between 1900 and 2010, broken down into border arcs, can be seen in Graph 09. It can be seen that the Southern Arc had a more significant population growth than the other two arcs from 1900 to 2010. It is the most populous, with more than 6 million inhabitants, which represents 58.66% of the total current border population. The Northern Arc has a population of approximately 2 million, or 18.29% of the border population, and finally the Central Arc with almost 2.5 million inhabitants, or 23.05%.

With regard to the spatial distribution of the border population, it can be seen that most of the population is located in the state of Rio Grande do Sul, which is explained by the fact that a large part of the territory of Rio Grande do Sul lies within the border strip and borders two countries: Uruguay and Argentina. The second largest state in terms of population is Paraná and the third is Mato Grosso do Sul, according to the 2010 demographic census. Graph 11 shows the distribution of the border population by state.

---

[36] Growth rates were calculated according to the following formula:

$$\sqrt[n]{\frac{P_{(t+n)}}{P_{(t)}}}$$

where P(t+n) and P(t) are the populations corresponding to two dates and n is the time interval between these dates (IBGE, 2014). Available at:
http://www.ibge.gov.br/home/estatistica/populacao/condicaodevida/indicadoresminimos/conc eitos.shtm

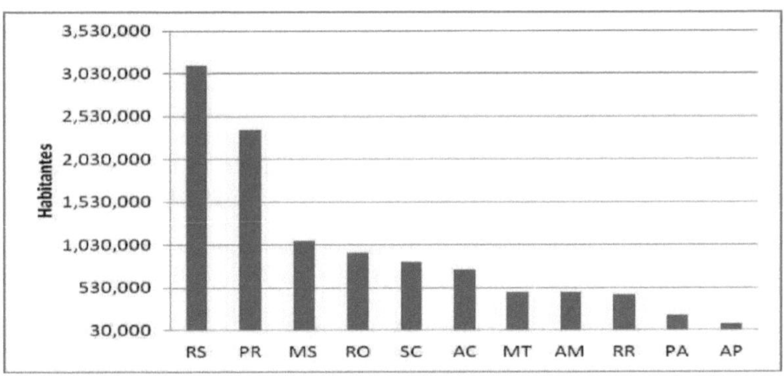

GRAPH 11 - DISTRIBUTION OF THE BORDER POPULATION IN 2010 BY STATE
SOURCE: IBGE. Own elaboration

The urbanization rate was 24% in the 1970s. Today, the urbanization rate is 75%, in other words, it tripled between 1970 and 2010. In comparison, the national urbanization rate in 1970 was 55.92% and today it is 84.36%. The urbanization rate in 2010 according to the border arcs can be seen in Table 08.

TABLE 08- URBANIZATION RATE IN 2010 ACCORDING TO BORDER ARCHES

Arches	Urbanization rate
Northern Arc	68%
Central Arch	78%
Southern Arc	77%

Source: IBGE. Own elaboration

The rate of urbanization on the border strip has urban differences between the arcs. The highest rate of urbanization is in the Central Arc, followed by the Southern Arc and the Northern Arc. Therefore, the phenomenon of urbanization does not occur in a homogeneous and generalized way on the border strip.

4.9 Urban hierarchy of border cities

The notion of hierarchy is related to the analysis of inequalities in weight and size between geographical objects, in a theory in which relationships are explained by interactions between them. The differences in the size of these objects appear very unevenly.

The most frequent statistical form of the distribution is a pyramid, in which the base is very numerous with small objects, has an average number of medium-sized objects and a very small number of large ones. It should be emphasized that the level of hierarchy in geography does not mean a hierarchy of power, political or administrative. Hierarchical relationships can often exist and be formed by unequal weights between objects (SAINT JULIEN; PUMAIN, 2010).

The finding of inequalities in size and weights prompts research into whether they accompany other, more quantitative differences, which often reveal the existence of various levels of complexity between systems (SAINT JULIEN; PUMAIN, 2010).

The size of cities is the product of a long-term process of local accumulation. The number of cities follows a geometric progression inverse to their size, described by a *rank-size rule,* in which cities are organized in descending order of size described by the equation log P = K-q log R, where P represents the population of a city, R its size in the urban hierarchy and K is a constant. Zipf[37] formulated the Pareto model in the form of the above equation, in which the population P of a city i is ordered by decreasing size according to its position in a national or regional set of cities (SAINT JULIEN; PUMAIN, 2010). The distribution of the size of cities also follows a certain regularity of their arrangement in geographical space (PUMAIN, 1997), since the distribution of cities is not disordered.

In the case of border municipalities, applying the *rang taille* law, the first places in the *ranking* in order of importance are occupied by Porto Velho (RO), Rio Branco (AC), Pelotas (RS), Cascavel (PR), Boa Vista (RR) and Foz do Iguaçu (PR), municipalities with more than 200,000 inhabitants. Thus, among the most populous municipalities, there are representatives of the three border arcs, but they are not numerous.

Graph 12 shows the *rang taille* curve, which indicates that there are few large cities on the border strip. And the larger cities are not big enough to be on the trend curve, justified by their low demographics compared to other regions. The trend curve shows that the number of small municipalities follows this curve, being the "pyramidal" base of the urban hierarchy on the border. The main result of the rang-taille curve is that the border municipalities do not form a separate urban network.

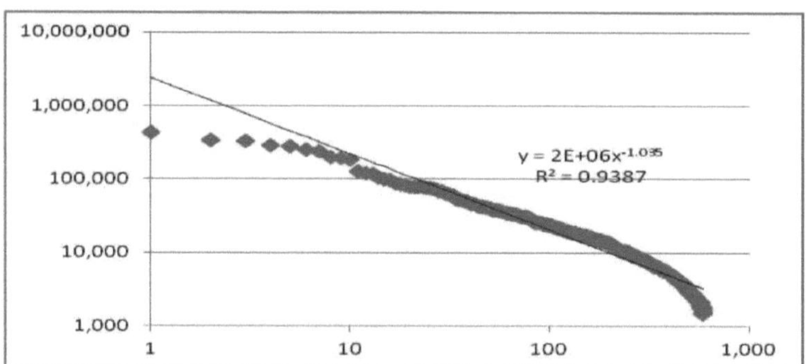

GRAPH 12 - *RANG TAILLE* CURVE OF BORDER MUNICIPALITIES
SOURCE: IBGE. 2010 Census. Own elaboration

When analyzing the three arcs together, in the *ranking of the* first 10 hierarchical positions, the municipalities of the Southern Arc predominate (6 positions out of 10), as the population is concentrated in this region. It is important to note that the first position is occupied by Porto Velho (RO), so, considering the entire border strip, the municipality with the largest population is located in the state of Rondônia. The next municipality in terms of population in the state of Rondônia is Vilhena, in 25th place[a].

The municipalities with populations between 100,000 and 200,000 are: Rio Grande (RS), Dourados (MS), Chapecó (SC), Uruguaiana (RS), Toledo (PR), Bagé (RS) and Corumbá (MS). Segmenting by arc, the Southern Arc lists the cities with more than 100,000 inhabitants in 2010, in descending order: Pelotas, Cascavel, Foz do Iguaçu, Rio Grande, Chapecó, Uruguaiana, Toledo, Bagé and Umuarama. In the Central Arc we have Porto Velho, Dourados and Corumbá and in the North Arc, Rio Branco and Boa Vista, which are state capitals. The other cities with more than 100,000 inhabitants are classified by the IBGE as regional or sub-regional capitals. A similar situation occurs

---

[37] G. K. Zipt was the first to systematically study the distribution of city sizes in various countries and at different times ((SAINT JULIEN; PUMAIN, 2001).

in the Northern Arc, where the largest population is in the municipality of Rio Branco, the most important municipality in the arc in demographic terms.

According to the population distribution by city size, 57% of those located on the border strip are municipalities of up to 10,000 inhabitants, 36% between 10 and 50,000 inhabitants, 3% between 50 and 100,000 inhabitants and only 2% have a population of over 100,000 inhabitants. Figures 11, 12, 13 and 14 show the location of border municipalities according to population size. The distribution by border arc shows that the Northern and Central Arcs concentrate municipalities with between 10 and 50 thousand inhabitants and the Southern Arc concentrates towns with up to 10 thousand inhabitants.

FIGURE 11 - BORDER MUNICIPALITIES WITH UP TO 10,000 INHABITANTS
SOURCE: from the author

FIGURE 12 - BORDER MUNICIPALITIES WITH 10 TO 50 THOUSAND INHABITANTS
SOURCE: from the author

FIGURE 13- BORDER MUNICIPALITIES WITH 50 TO 100 THOUSAND INHABITANTS
SOURCE: from the author

FIGURE 14 - BORDER MUNICIPALITIES WITH MORE THAN 100 THOUSAND
INHABITANTS
SOURCE: from the author

The Northern Arc, the least populous of the three, has a total population of 1.9 million people, 68% of whom are located in cities. In terms of urban hierarchy, Rio Branco is the most populous, performing central administrative functions because it is the capital of the state of Acre, followed by Boa Vista, capital of the state of Roraima and Cruzeiro do Sul, which is not a capital but has a reasonable population relative to the other border cities of the Northern Arc, although in comparison with the population of Boa Vista (RR), Cruzeiro do Sul (AC) is 3.6 times smaller. Following in descending order of importance in the demographic hierarchy are: Oriximiná (PA), Alenquer (PA), Tabatinga (AM), Óbidos (PA), Laranjal (AP), Sena Madureira (AP) and Sao Gabriel da Cachoeira (AM).

From the point of view of state analysis, it can be seen that the state with the largest border towns in the Northern Arc is Acre, which is partly justified because the entire territory of Acre is located on the border strip. Interestingly, the state of Pará, despite having few border municipalities, has an interesting demographic aspect: Oriximiná *ranks* fourth in the Arco Norte hierarchy and the state ranks seventh with Óbidos. It is the state with the fewest border municipalities, located close to the border with Suriname.

The population gradient varies between 3,700 inhabitants and 163,000 inhabitants, explained by the functions carried out in each municipality. The smallest municipal border population in Arco Norte is in Pracuúba (AP), with 3,793 inhabitants, and the closest border is with French Guiana. The urbanization rate in this municipality is 49%, i.e. the rural population is larger than the urban population.

The Central Arc has a population of approximately 2.4 million, slightly more populous than the Northern Arc, and the urbanization rate is higher, with 78% of the population living in cities. In terms of the demographic hierarchy applied only to the Central Arc, the municipality of Porto Velho, which is also the state capital, is in first place, followed by Dourados, which is the regional capital, and Corumbá (MS), which in comparative terms has % of the population of Porto Velho. Following the hierarchy are Cáceres (MT), Tangará da Serra (MT), Ponta Pora (MS), Vilhena (RO), Rolim de Moura (RO), Naviraí (MS) and Aquidauana (MS). Between Porto Velho (RO) and Dourados (MS), the population of Dourados is 2.2 smaller than Porto Velho. Among the top 10 positions, 5 municipalities are from Mato Grosso.

Porto Velho and Corumbá are located close to the Bolivian border and Dourados, close to the border with Paraguay. The population ranges from 428,527 inhabitants in Porto Velho (RO) to 2,315 inhabitants in Pimenteiras do Oeste (RO), located on the border with Bolivia.

The Southern Arc, the most populous of the three, has a total population of 6.3 million people, accounting for 59% of the border population, i.e. more than half of the border population lives in the Southern Arc. Despite this, the urbanization rate is 77%, not significantly different from the Central Arc, which has a rate of 78%, but different from the Northern Arc, which has an urbanization rate of 68%.

In terms of the demographic hierarchy, the municipality of Pelotas (RS) has 328,000 inhabitants, followed by Cascavel (PR) with 286,000 inhabitants and Foz do Iguaçu with 256,000 inhabitants. Following in descending order of importance in the demographic hierarchy are: Rio Grande (RS), Chapecó (SC), Uruguaiana (RS), Toledo (PR), Bagé (RS), Umuarama (PR) and Erechim (RS).

From the point of view of state analysis, the only city in Santa Catarina in the top 10 of the demographic hierarchy is Chapecó, which is partly justified because Rio Grande do Sul and Paraná have a larger number of border municipalities than Santa Catarina. The second largest border population in Santa Catarina is Concórdia, with 68,000 inhabitants, in eighteenth place among the Southern Arc cities.

The population gradient of the Southern Arc varies between 328,000 inhabitants and 1,400 inhabitants, explained by the functions carried out in each municipality. The smallest border municipality in the Southern Arc is Sao Tiago do Sul, with 1,465 inhabitants, and the closest border is with Argentina. The urbanization rate in this municipality is 44% and the rural population is larger than the urban population; it is also the smallest municipality among the arcs.

The urban hierarchy changes over time. Comparing the 2000 and 2010 censuses, it can be seen that among the top 10 places in the *ranking* by order of importance, the first, seventh, eighth, ninth and tenth positions continue to be occupied by the same cities in the two consecutive censuses: Porto Velho, Rio Grande, Dourados and Chapecó. The cities that lost position were Pelotas and Foz do Iguaçu. In 2000 they were second and third respectively, and in 2010 they were third (Pelotas) and sixth (Foz do Iguaçu). In the opposite direction, there is Rio Branco, which went from fourth to second position, Cascavel, which migrated from fifth to fourth position, and Boa Vista, from sixth to fifth position in the *ranking*.

Part 3
Chapter 5

5 ECONOMIC INTEGRATION THEORY AND THE PERSPECTIVE
SOUTH AMERICAN

> Contrary to superficial assertions, it is not just any integration that is a desirable goal, but external integration that strengthens and develops internal links.
>
> **Rubens Ricupero**

This chapter aims to demonstrate the context of economic integration in the Southern Cone by briefly outlining the history of Latin American integration and the changes in Brazilian foreign policy, from the import substitution model to open regionalism. The changes in territorial policy are then discussed from a South American perspective, highlighting the main economic blocs and MERCOSUR's foreign trade relations. This chapter is justified in order for the reader to understand the context of the international scale that directly influences trade flows in the municipalities studied.

5.1 BACKGROUND TO INTEGRATION POLICY IN BRAZIL

To briefly recap the predecessors of the integration policy, there were different moments in Brazilian economic history. In the 1930s, after the international crisis of 1929, the federal government implemented the Import Substitution Policy (PSI). Gremand, Vasconcelos and Toneto Junior (2007) state that the main characteristic of the PSI is closed industrialization, due to two elements:

- be inward-looking, i.e. aim to serve the domestic market, not be an industrialization that produces for export;
- depend largely on measures that protect the domestic industry from foreign competitors.

The world economy then experienced the greatest economic recession of the 20th century. After the crash of the New York Stock Exchange, European countries and the United States applied extremely protectionist policies, which led to a reduction in trade, the imposition of customs barriers, quotas and the exaltation of nationalism - what prevailed was what was *made in* Europe.

Under the paradigm of national developmentalism, the PSI was an important instrument for the industrialization and modernization of the country, aimed at developing the manufacturing sector and resolving dependence on foreign capital. Within this context, the import of capital goods was vital for the recovery of the Brazilian economy, the government controlled import rates through quotas and the exchange rate was generally unfavorable for importing products.

To clarify, the authors Gremaud, Vasconcelos and Toneto Junior (2007) explain how the PSI cycle works:

a) the beginning of an external bottleneck, generating a shortage of foreign currency;

b) the government attempted to control the currency crisis by means of measures that made imports more difficult in order to protect domestic industry;

1. generating a wave of investment in import-substituting sectors, increasing national income and aggregate demand;

2. a new external bottleneck due to the growth in demand itself (back to point 1).

The PSI was defended by the Economic Commission for Latin America and the Caribbean (ECLAC), which stated that if Latin American countries continued to export primary products and import industrialized products, the terms of trade would deteriorate. This would push the countries of the region further and further towards the bottom of the world economy because most nations would not be able to increase the portfolio of products exported from their territory.

Although the JK years were beneficial for national industry, due to the diversification and increase in industrial productivity, they were not as important for foreign trade. The export agenda was still basically agricultural, without adding value to the goods traded.

In the mid-1960s, the Brazilian government realized that it could export products with higher added value. As the industrial park was already mature, the country was able to launch itself onto the international market. Foreign trade came to be seen as a development mechanism. The surplus of consumer goods produced for the middle class was absorbed by a policy of encouraging exports.

In the 1970s, after the first oil shock, Brazil maintained its growth rate through industrial investment and foreign debt. Combining the stimulus of investment, the promotion of exports and the restriction of imports, Brazil managed to maintain its growth until the end of the decade.

The 1980s, known as the lost decade, was a turbulent period for the Brazilian economy. The context was one of high inflation, foreign debt and low economic growth. Internationally, the world was divided into two ideological blocs: capitalist and socialist.

The end of the Cold War had a variety of meanings for Latin America, undoubtedly the most important of which was the reorientation of its foreign policy. The new unipolar world order and the process of re-democratization of countries are key factors in the new economic issues. Known as the Washington Consensus, the wave of economic liberalism dictated the rules for Latin countries.

The neoliberal decalogue consisted of: fiscal discipline, trade openness, tax reform, privatization of state-owned companies, market exchange, elimination of restrictions on foreign direct investment (FDI) and property rights (MDIC, 2010).

As you can see, the policy was based on economic liberalism, i.e. the law of the market was the valid rule. The government could not intervene in the commercial relationship of the market, the law of supply and demand would prevail in the commercial relationships between the parties.

Integration policy emerged in this context. Since the 1990s, globalization has brought economic changes. The search for privileged trade relations between countries led to new power relations emerging in the economic sphere.

7.2 ECONOMIC INTEGRATION THEORY

The emergence of economic blocs has undoubtedly been driven by globalization, which has brought about a series of transformations in the world economy. The issue of regional integration - in its commercial, productive and financial aspects - has been discussed in various contexts in South America and by different international trade organizations. Despite historical and political efforts, which have not always been successful, the integration of South American markets is taking place slowly and gradually.

In the early 1990s, at the height of liberal theories, it was believed that economic interests would predominate in international relations, due to the end of political and military polarization. A new moment began, with discussions on economic integration theory intensifying, the ideas of bloc formation, tariff unification, globalization and trade flows being discussed in academia and governments.

The wave of regionalism represented an attempt by members to facilitate participation in the world economy, rather than isolating themselves. These regional processes involved both developed and developing countries, and can be characterized as strategies to liberalize and open up economies by implementing *export* and *foreign-investment-led* policies, rather than promoting import substitution strategies as had occurred in Latin America (SENHORAS, 2007).

The significant increase in intra-regional trade during the 1990s showed that there was no reduction in trade between economic blocs; what happened was precisely the opposite, an increase in international trade through regionalism and bi- and multilateral trade agreements (SENHORAS, 2007).

The phenomenon of regionalization became pulsating, and politics and economics began to be organized in such a way as to achieve bargaining power between states. Regional schemes were also a way of reconfiguring the geopolitical distribution of power, as well as bringing new territorial configurations due to the formation of new economic blocs.

The paradigm for economic integration sought non-discrimination between countries, and even the most disadvantaged nation would be treated equally. The World Trade Organization[38] was created with the responsibility of "adjusting" possible disagreements between nations.

Economic integration has led to the inclusion of new authors, new powers and a new reality

---

[38] The WTO originated from the General Agreement on Tariffs and Trade (GATT).

in the territories of the states. Authors such as Moreira, Ernst and Mitrany, Keoany and Karl Deustch[39] discuss the theorization of European integration, but this will not be the focus of this work. In Brazil, Alcides Vaz is dedicated to the study of integration in South America, focusing mainly on MERCOSUR. There are other Brazilian researchers who contribute to academia, such as Paulo Almeida, Alfredo da Mota Menezes, Pio Penna Filho, Wanderley Costa, among others.

According to diplomat and researcher Paulo Almeida, the concept of economic integration applies to entities of different political natures, with different economic realities, but it is best understood if it is considered as a process in successive stages: 1) *area of tariff preferences*, which involves the simple selective reduction of tariffs between two or more members, without complementary obligations as a commercial policy; 2) *free trade area*, which completely liberalizes trade between members within a given period, while each retains its own tariff structure in relation to third countries; 3) *customs union*, which also includes the definition of a common external tariff; 4) *common market*, which completely liberalizes the flow of productive factors and people, as well as obliging the adoption of common policies in the commercial, industrial, agricultural and competition areas, among others; and finally, 5) *economic and monetary union*, which may include, as in the case of the European Union (EU), the abolition of national currencies in favour of a common circulating medium for its members (ALMEIDA, 2002).

The geographer Lobo (1997) classifies those that deepen integration more succinctly:

a) free trade area, with the abolition of tariffs and quantitative restrictions between the participating countries, each maintaining its own tariff with regard to third countries;

b) customs union, abolishing discrimination in the movement of goods within the union, and establishing common tariffs for the import of goods from other countries;

c) economic union, combining the removal of restrictions on the movement of goods and discriminatory factors with a certain harmonization of national economic guidelines;

d) full economic integration,[40] with the unification of monetary, fiscal and social policies, and the obedience of member states to a supranational authority in economic matters.

Researchers Menezes and Penna Filho complement the idea put forward by Lobo. The first step would be a preferential tariff agreement between certain countries. Tariffs between the members involved would be lower than those charged to other countries not participating in the integration. Customs barriers would be reduced or eliminated for trade within the integrated zone. The difference here would be that the countries signing the agreement would not adopt a common external tariff, which would be applied equally to non-members of the integration. Trade would be open and exchanges facilitated, but each country would maintain its own specific taxes for transactions with third parties outside the bloc. The customs union is a more advanced step and is a little more complicated, especially between developing countries. In addition to eliminating bureaucratic obstacles to trade between the participants, there is a common external tariff for the integrated countries, to be applied equally to countries outside the bloc (MENEZES; PENNA FILHO, 2006, p. 2).

Moreira (2003) states that for integration to be successful it needs to take place in three sectors: economic, social and political. Two instruments can be used to develop economic power between the parties: eliminating customs barriers between member states, allowing the free movement of goods and capital, and defining a single common economic policy towards states outside the common market.

The political sector is the most difficult. The historical values of patriotism, loyalty to one's homeland, nation and state, national identities and sovereign independence are all issues that need to be addressed. This is why integration has as its corollary the tolerance and cooperation of differences under the same political institution. The process of setting up transnational authorities and the transfer of loyalty are the most delicate areas (MOREIRA, 2003).

---

[39] He is one of the leading theorists on International Relations.

[40] The European Union is the best example of this type of economic bloc. It adopts a single economic policy and currency.

Alcides Vaz (2002) shows that the definition of integration is directly related to the promotion of states' security interests, and is conceived as a strategy aimed at controlling and, if possible, suppressing conflicts at regional level. A process of integration at the international level is above all consensual, or communitarian, based mainly on the development of shared norms, values, interests or objectives.

Vaz doesn't come up with a new theory of integration, he just analyzes existing theories and looks at MERCOSUR in particular.

Mercosur integration differs from other blocs such as the European Union in that it is not supranational. The MERCOSUR issue will be detailed later in this chapter.

The authors Menezes and Penna Filho (2006) argue that for some, economic integration occurs when the prices of all the same products are equalized in a given region. There would be a single market where the same price would be charged for the same good. For other authors, integration is simply the elimination of economic barriers between two or more economies. A third understanding sees integration as the result of the elimination of all impediments to international trade between countries, with some general coordination mechanisms in the integrated economies.

The main sign of integration between countries would be the absence of customs and tax offices between the integrated partners. Others argue that economic integration is nothing more than the division of labor in a region (MENEZES; PENNA FILHO, 2006). These authors have a strictly economic focus.

In the opinion of Celso Lafer[41] , integration presupposes the elimination of barriers to trade and the drawing up of common, or at least compatible, rules on trade-related issues, plus access to markets; compatible rules are necessary but not sufficient conditions. A third and no less important aspect is the physical infrastructure of integration. The effective occupation of national spaces requires transforming the classic separation borders into modern cooperation borders, in order to make the development of border regions that have long been isolated corners economically viable. A fourth element, which becomes more visible in times of financial turmoil, but which has permanent relevance, is that of macroeconomic coordination. In addition to the issue of synchronizing national economic cycles, emerging countries are more vulnerable to fluctuations in the rate of growth of the world economy, the prices of raw materials, investment flows and the liquidity of foreign exchange markets. These factors must occur simultaneously (LAFER, 2002).

For Paulo Almeida, the concept of integration is eminently positive and proposes a new reality of unrestricted cooperation and the construction of solid economies and more inclusive societies (ALMEIDA, 2006). The theory of integration can be classified into three levels, according to the geographical scope between two or more communities: national, regional and global. Whatever degree of integration is considered, it must imply the existence of conditions that make it possible to resolve the various conflicts inherent in the process.

The advantages of economic integration include:

- enables the economies of scale necessary for companies to reduce costs;
- facilitates the circulation of the essential factors for production: raw materials, labor and capital;
- favors access to consumers and energy sources;
- creates a broader environment than the nation state, in which economic agents can get used to external competition, but still have institutional safeguards and bulkheads.

Other advantages of economic integration, according to Morini and Simoes (2002), are the following:

- general increase in production;
- increasing productivity by exploiting comparative advantages between members of the same economic bloc;

---

[41]Celso Lafer was Minister of Development, Industry and Trade and Minister of Foreign Affairs on two occasions, in 1992 and from 2001 to 2002, under Fernando Collor and Fernando Henrique Cardoso, as well as Brazil's ambassador to the WTO, and Brazil's ambassador to the United Nations (UN) from 1995 to 1998.

- stimulating efficiency due to increased internal competition.

In static terms, a process of economic integration allows for an increase in the level of productivity of the countries that make up the bloc, while in dynamic terms, it produces an increase in the rate of growth of this productivity. Also in dynamic terms, regional integration allows for gains in the learning process and technological innovation. A larger market, providing greater specialization and amortization of technological investments, tends to accelerate innovation, creating agglomeration economies and *spillover* effects or linkages in production chains (DATHEIN, 2007).

Brazilian geographer Wanderley Messias da Costa says that the ongoing process of integration in South America has been repeatedly defended in order to strengthen the country's presence in the subcontinent as a whole. Betting on its integration is even a *sine qua non* condition for any other negotiations of this kind on another scale (COSTA, 1999).

The degree of integration between different regions is indicated by the structure of inter-regional flows of goods and services, in which an economy will be more integrated the greater its propensity to import from the other region. Thus, an increase in demand for the region's products will increase its demand for goods and services from the area itself, from other regions and from abroad (MARIZ; ZEBRAL FILHO, 1998).

The so-called key regions in the generation of production/employment are those that simultaneously exert vertical (purchase of goods and services) and horizontal (sales) linkage effects above the system average. The long-term reproduction of this mechanism will make the productive structure of each region more homogeneous, reducing regional inequalities. In addition, the economic integration and internal diversification of each region will help to ensure that external crises do not have such an intense impact, as their effects are more evenly distributed (MARIZ; ZEBRAL FILHO, 1998).

Although the idea of integration is "seductive," it is not "perfect". It is a difficult process to negotiate because trading partners are not always willing to give up tariff advantages. Negotiation is complex because there are many divergent and contradictory interests, so the process is extremely slow and gradual.

Although the process is positive for international bargaining and for increasing the quality of products encouraged by competition, integration can also bring negative elements during its course, such as the bankruptcy of companies that are not so adaptable, the invasion of products manufactured outside the national territory, pressure on the labor market, etc.

Economic integration produces new and positive opportunities, but also threats to social and economic stability. In regional, sectoral and labor market terms, obsolescence can occur, imposing transition and adjustment costs with income redistribution. For this reason, there must be reward and retraining policies. Furthermore, integration exposes asymmetries in productivity which mean opportunities for growth, but also threats deriving from social and economic balances (DATHEIN, 2007).

7.3 BRAZILIAN TERRITORIAL POLICY IN THE CONTEXT OF SOUTH AMERICAN INTEGRATION

The so-called "classic" regionalization of Latin America is based on geomorphological macro-compartmentalization applied to regional analysis. Traditionally, the terms Platinum Basin, Amazon Basin and Andean Basin were used (COSTA, 1999).

Over time, Brazilian territorial policies have undergone modifications due to changes in government paradigms. Regional analysis has coexisted with a number of regional development paradigms, such as the Developmental State, the Liberal State and, currently, the Logistical State.[42]

Regionalism in Latin America was influenced by the European Coal and Support Community (ECSC), which served as an inspiration. The ECSC, which later became the European Union (EU), exported a successful model, although the socio-economic characteristics between the Latin and European blocs were very different.

The *Asociación Latinoamericana de Libre Comercio* (ALALC) was created in 1960 by the

---

[42] Classification used in International Relations. For more details, see "Brazil's International Insertion", by Amado Cervo.

Treaty of Montevideo, based on the most favored nation clause and the principle of national treatment, to create a free trade area within twelve years. The initiative was unsuccessful because the proposal was rigid and very ambitious, it didn't take into account the economic and industrial heterogeneity of the region, and it didn't have a dispute settlement mechanism that could guarantee the agreements made (BARRAL; BOHREL, 2010). There was a mismatch between the totalizing agenda of the integration project and the advance of protectionism in the countries of the region, in a national-developmentalist context, both in democratic and authoritarian periods.

In the 1980s, the Latin American Integration Association (ALADI) replaced the ALALC and inherited its characteristics. The Cartagena agreement provided for trade liberalization, coordination of industrial development policy, special treatment for multinationals, unified import programs, creation of a development corporation, joint action in scientific research and education, creation of a common external tariff, coordination and harmonization of their economic policies and investments in infrastructure (ALADI, 2013).

The original ALADI countries are: Argentina, Bolivia, Brazil, Chile, Colombia, Ecuador, Mexico, Panama, Paraguay, Uruguay and Peru,
Venezuela and Bolivia. Cuba joined in 1998 and became a full member in 1999. The same happened with Panama, which applied for membership in 2009 and became a full member in 2012. Nicaragua's accession process is still underway (ALADI, 2013).

The Andean Community (CAN), created in May 1969 by the Andean Pact signed by Bolivia, Chile, Colombia, Ecuador and Peru, established integration between these economies. There were disagreements with the ALADI due to the lack of gains by the smaller economies (MENEZES; PENNA FILHO, 2006) and the CAN formed a bloc with common goals and objectives, setting up a liberation program and establishing a common tariff between the members.

The exhaustion of the import substitution model and the advance of neoliberal strategies in the 1990s imposed a new agenda for Brazilian foreign policy, a process that began with the economic opening of the Fernando Collor de Mello government (1990-1992), which provoked reforms and resistance. **The period was marked by the expansion of regional economic blocs around the world.** The Southern Common Market (MERCOSUR) was born as part of this phenomenon.

MERCOSUR was signed in 1991 by Argentina, Brazil, Paraguay and Uruguay in the Treaty of Assunpao. The bloc was formed as a result of geopolitical stability in the region, which today can be seen in the reduction of border disputes or disputes of any other nature to practically negligible levels. What has also contributed to the end of rivalries and hostility is the consolidation of democratic regimes in a region historically marked by political instability. In addition, the construction of multilateral cooperation arrangements offers effective conditions for building the institutional framework required by integration movements (COSTA, 1999).

Since South America has historically been the priority regional area for exports of Brazilian industrialized products, regional integration has become a pressing need as a way of protecting industrial sectors from fierce competition from other countries.

With regard to the subcontinent of South America, it is only now that we can see more emphatic moves by Brazil in the direction of explicitly integrationist territorial policies, capable of involving the wide range of countries with which it has so far shared little in the process of economic, social, political and cultural development. The country is implementing a strategy adopted more than a decade ago, with developments in foreign policy for the region, making full use of its comparative and competitive advantages (territorial configuration, position, leadership and diversified productive structure), making itself the driving force behind this integration process (COSTA, 1999b).

Another project aimed at integration, in a broader sense, is the Initiative for the Integration of South American Regional Infrastructure (IIRSA), which emerged at the meeting of 12 South American presidents in Brasilia on August 31, 2000. The countries that signed the agreement were: Brazil[43] , Argentina, Bolivia, Chile, Colombia, Ecuador, Guyana, Paraguay, Peru, Suriname, Uruguay

---

[43] On this occasion, the president who took over IIRSA was Fernando Henrique Cardoso, although President Luiz Inácio Lula da Silva used the initiative as a political platform.

and Venezuela. The presidents agreed to carry out joint actions to boost the process of political, economic and social integration, including the modernization of regional infrastructure and specific actions to stimulate the integration and development of the sub-regions (IRSA, 2013).

During the 2000 summit, issues relating to democracy, illicit drugs and related crimes, trade integration, information, technology and knowledge, and the integration of physical infrastructure were discussed (COUTO, 2009). With the signing of IIRSA, the South American presidents committed themselves to investing in physical infrastructure, communications and energy in the Southern Cone. The agreement provides for major investments in these areas and directly impacts the border regions.

The document includes multilateral mechanisms for financing the work, such as the Inter-American Development Bank (IDB), the Development Bank of Latin America (CAF) and the Financial Fund for the Development of the River Plate Basin (FONPLATA).

From a geographical point of view, IIRSA is an initiative whose guiding principle is open regionalism, with the space divided into geo-economic regions, forming integrated development axes, the EIDs. They were drawn up in multinational territories where the population, productive areas and existing trade flows are concentrated. The EID aims to boost the region's productive potential and provide access to areas that are currently isolated due to poor logistics, telecommunications and energy. The axes mapped out cover the whole of South America and are called: Amazon Axis, Andean Axis, Southern Andean Axis, Capricorn Axis,[44] Axis Guiana Shield, Paraguay-Paraná Waterway Hub[45] , Central Interoceanic Hub, Mercosur-Chile Hub[46] , Peru-Brazil-Bolivia Hub, Southern Hub (Figure 15).

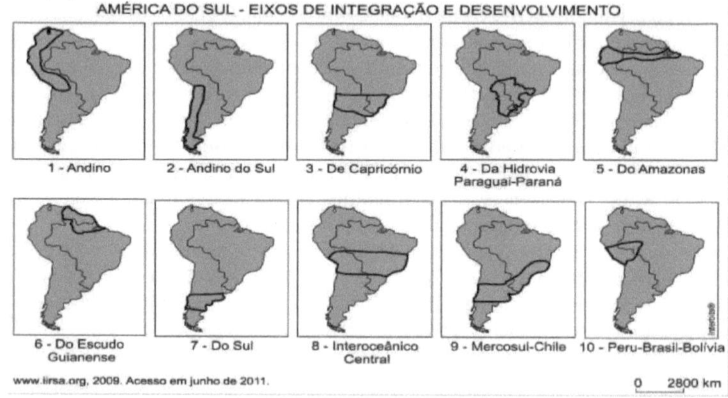

FIGURE 15 - IIRSA AXES
SOURCE: https://professorjamesonnig.wordpress.eom/2012/11/13/fuvest- comentada-2/

All the axes make up transportation, telecommunications and energy projects and are strategic areas for regional development. It's important to note that these projects are continental in scope. For the first time, there has been a supranational agreement involving so many countries, agreeing on common interests.

Another relevant bloc is the Union of South American Nations (UNASUR), which was created on the initiative of Brazilian diplomacy and has international legal personality[47] . Initially called Comunidad Suramericana de Naciones, it was born in 2004 at the III Sudamerican Meeting of Presidents. On April 16, 2007, the name was changed. UNASUR was also an instrument of the Lula

---

[44]Bioceanic corridor between Chile and Brazil. Direct influence on Foz do Iguaçu as a port for the flow of grains.

[45] River development in the Río del Plata basin. Port influence in the twin cities of Rio Grande do Sul.

[46] Location of the twin cities on the border with Uruguay.

[47] Article 1 of the UNASUR Constitutive Treaty available at:
http://unasursg.org/PDFs/unasur/tratado-constitutivo/Tratado-Constitutivo-version- portugues.pdf

government and aims to improve regional development. Its action plan focuses on seven main priority areas:

    a. political dialog,
    b. physical integration,[48]
    c. environment,
    d. energy integration,
    e. financial mechanisms,
    f. promoting social cohesion, social inclusion and social justice,
    g. telecoms.

The proposal put forward is not only for trade cooperation and economic complementation, but also for the integration of the transport network, such as the transoceanic Atlantic-Pacific road connection in the Amazon region and rail in the Platinum region, with Chilean ports, as well as waterways between the Amazon, Platinum and Caribbean basins (Orinoco River). In addition, there would be energy integration and the political-diplomatic cooperation. This initiative stems from the projection of the Brazilian economy and the country's diplomacy (UNASUR, 2012).

7.4 MERCOSUR

The starting point for MERCOSUR was the signing of the Treaty of Accession in 1991 by the States Parties: Brazil, Argentina, Paraguay and Uruguay. The preparations for MERCOSUR began in 1988 with the signing of the Treaty of Integration, Cooperation and Development between Brazil and Argentina. Both countries were going through a period of economic liberalism and democratization. This treaty was important in creating synergy for the formation of the future economic bloc. The neighbors took on characteristics that transcended this spatial circumstance, including the sharing of production chains, increased commercial density, political and legislative harmonization.

In 1990, the Buenos Aires Act was signed, referring to the primary MERCOSUR negotiations between Brazil and Argentina. A year later, Uruguay and Paraguay joined the negotiating round and the Treaty of Assunpao was signed by all. The Treaty provided for the implementation of the Southern Common Market as of December 31, 1994, through a trade liberation program with linear and automatic tariff reductions. Non-tariff restrictions would also be gradually eliminated.

MERCOSUR's golden age was between 1991 and 1998 (PRAZERES, 2006). Intra-bloc exports and imports increased considerably. In this scenario, South America emerged as a privileged area for Brazilian foreign policy to increase exports (Graph 13).

During this period, Argentina, the group's main trading partner, was going through a privatization phase, and Brazil was going through a period of political instability due to the *impeachment* of President Fernando Collor, but MERCOSUR maintained a good trade relationship with its members. It increased the production of national industry because it found in neighboring countries a favorable market for the purchase of Brazilian products. Argentina became Brazil's second most important trading partner.

In 1994, the countries signed the Ouro Preto Protocol, which expanded MERCOSUR's institutional structure by giving it legal personality under international law. Since then, MERCOSUR has been made up of the Common Market Council (CMC); the Common Market Group (GMC); the MERCOSUR Trade Commission (CCM); the Joint Parliamentary Commission (CPC); the Economic and Social Consultative Forum (FCES) and the MERCOSUR Administrative Secretariat (SAM).

MERCOSUR is currently classified as a customs union, i.e. as trade integration not only from the point of view of free trade (gradual end of tariff barriers), but also with the adoption of a common external tariff (TEC), which was due to be implemented in 1997. In practice, the TEC exists, but there is also a list of exceptions. In this way, MERCOSUR is considered an incomplete or imperfect customs union because the TEC presents a list of exceptions to the liberalization program for products

---

[48]This investment directly affects the twin cities because they will receive funds to improve the logistics network between the cities of the Southern Cone.

considered sensitive.

From 1999 to 2001, Argentina went through a period of severe crisis. Due to the crisis in Asia (1997) and Russia (1998), Argentina had a very strong sectoral deficit, which led the country to a high degree of fiscal and commercial volatility. What everyone feared happened: the Eastern crisis affected the reliability of investors in Latin America because economic interdependence shook structures. Brazil and Argentina adopted an exchange rate anchor as a way of combating the crisis. Combined with inflation and investor flight, the result was disastrous. In order to keep their currencies pegged to the dollar, they needed a constant flow of foreign currency, which was interrupted by the Asian crisis (ROCHA, 2006). As a result, in 1998 Brazil resorted to a foreign loan.

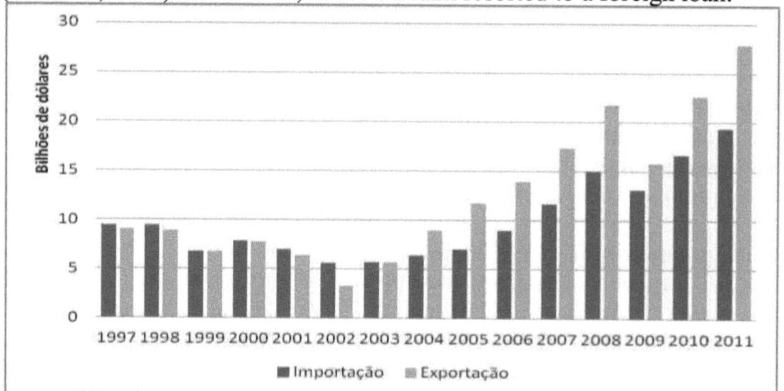

GRAPH 13 - INTRA-BLOC FOREIGN TRADE (MERCOSUR) - (US$-FOB)
SOURCE: MDIC/SECEX

President Fernando Henrique Cardoso's government decided to stabilize prices by devaluing the national currency and opening up trade, guaranteed by the entry and opening up of the capital account and stimulated by privatizations and the reform of the state. As a result, Brazil's exports became cheaper, which provoked complaints and concerns from Argentina's competitors. On the other hand, Argentina adopted the opposite policy, of overvaluing its currency, so Argentine foreign trade faced severe difficulties due to the inability to make adjustments to the exchange rate policy. The result was the 2001 crisis, possibly the most serious in Argentina's history. Consequently, the relationship with MERCOSUR was shaken. Argentina wanted the Customs Union to be transformed into a Free Trade Area because this would allow it to use protectionist resources to balance Argentina's trade balance (IPEA, 2010).

The period of crisis (Graph 14) highlighted the need to create efficient mechanisms so that each country did not act in isolation, without taking into account the repercussions of its decision on the neighboring member.

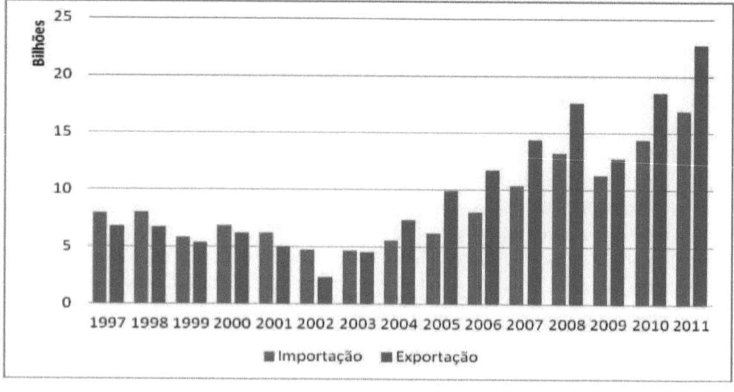

GRAPH 14 - BRAZIL'S FOREIGN TRADE WITH ARGENTINA - (US$-FOB)
SOURCE: MDIC/SECEX

The performance of MERCOSUR's two largest economies indicates that they are facing inverted cycles of economic growth. It should be noted that in seven years (1991-93, 1995, 1998, 1999 and 2000) this same pattern was observed. In three of them (1995, 1998 and 1999) this was directly related to instability in the international financial system and its consequences for the economies of both countries, while in the others it can be attributed to different domestic macroeconomic conditions and microeconomic imbalances (VAZ, 2001).

There was a return to growth in intra-regional trade flows from 2003 onwards, especially between the two major economies, but sectoral imbalances remained, motivated by the Argentine Industrial Union's demand for protection, generally met by the new administration of President Néstor Kirchner. Argentina believed that Brazil intended to reduce it to a mere role as a supplier of primary products, reserving for itself all the higher value-added chains, which, in a way, was confirmed in almost all areas, due to the tremendous effort of productive adaptation carried out by Brazilian industry in the course of the trade liberalization process of the 1990s and, later, due to the new gains in competitiveness acquired since the devaluation of 1999 (ALMEIDA, 2006). Since 2003, Brazil has accepted the imposition of restrictions on Brazilian exports of products such as footwear, through the mechanism of non-automatic licenses and quotas for sales of white goods - refrigerators, washing machines and fires - to the Argentine market (VIANA *et al.*, 2011).

The Argentine authorities repeatedly accused Brazil of unfair competition in attracting investment, thanks to tax incentives which added to the economies of scale of a market four times larger than Argentina's. This effect may have occurred in a concrete way, given its chain effects (ALMEIDA, 2006). This effect may have occurred concretely in the automotive sector, the essential basis of bilateral trade and a powerful factor in boosting growth given its chain effects (ALMEIDA, 2006).

Finally, at the beginning of 2006, both countries concluded Argentina's long-awaited project for a sectoral safeguards mechanism. The Competitive Adaptation Mechanism provides for the possibility of applying safeguards to bilateral trade. However, its use has been avoided, as it requires, among other things, proof of damage to the local industry, which has been enthusiastically welcomed by the Argentinians and widely complained about by the Brazilian industry (ALMEIDA, 2006).

The MERCOSUR Parliament and the Emergency Convergence Fund, aimed at reducing economic asymmetries within the bloc, were created. Finally, there was the conclusion of preferential tariff agreements with developing countries such as India and South Africa; trade agreements with Canada, Japan and Israel and the incorporation of two new associate members: Colombia and Ecuador. As for the relationship between Brazil and Argentina, mention should be made of the financing of government purchases and exports of Brazilian products with Argentine components through the National Development Bank (BNDES) and the Export Financing Program (PROEX) (MATHIAS; GUZZI; GIANINNI, 2008).

As far as the trade relationship between Brazil and Paraguay is concerned (Graph 15), MERCOSUR is not just a free option for the latter, but a necessity in the face of the new international reality, which requires the reformulation of action strategies in order to defend the country's interests and avoid the isolation of the state in the globalized world.

Among the Southern Cone countries, Paraguay is perhaps the one that poses the greatest challenge to regional integration and therefore deserves more attention from Brazil. Despite being one of the smallest members of MERCOSUR, Paraguay has "ills" that come together in various ways and with such intensity. Paraguay has indissoluble ties with Brazil and the resentment between the two is intertwined or directly related to issues on the bilateral agenda. Paraguay's dissatisfaction with the bloc, which has brought it little or no advantage in economic and political terms, is numerous. It is worth noting that Paraguay's trade with its MERCOSUR partners remains in deficit, with the exception of Uruguay (SILVA, 2007).

The process of trade opening brought about by MERCOSUR served to stimulate an increase in Brazilian exports, especially of products in which the country had a great comparative advantage.

The result was accelerated growth in Brazilian foreign trade. The main Brazilian products exported to Paraguay are: minerals, seeds, oleaginous fruits, meat, soy flour, soy oil, cereals and wood.

Paraguay is the country in the bloc with the lowest GDP, along with Uruguay. In political terms, it has little say within the bloc, which highlights Paraguayan dissatisfaction. In terms of Paraguay-Brazil physical integration, the borders are weak and are used by drug trafficking and smuggling networks, which has caused concern.

According to geographer Fernando de Lima, trade relations between Brazil and Paraguay grew 2.7 times over a 20-year period, from 1989 to 2009. However, this relationship was much more favorable for Brazil: the trade flow between the two countries grew at a geometric rate of 5% per year; however, exports increased by 6.4% per year, while imports grew by only 2.5% per year, which leads to a trade balance growing at a rate of 11.2% per year (LIMA, 2011).

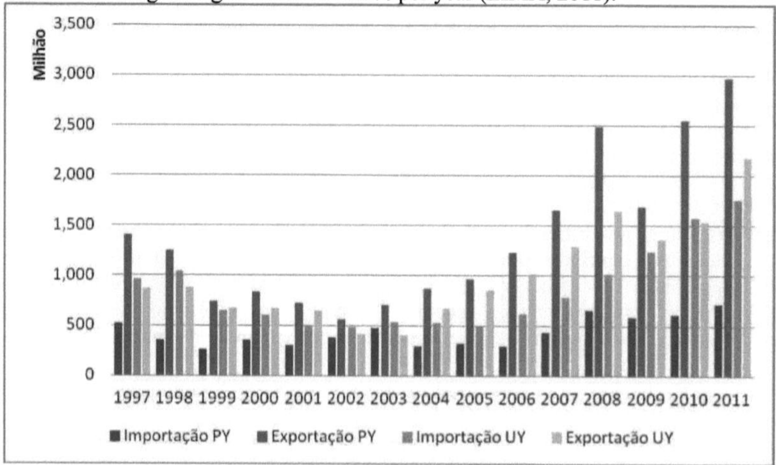

GRAPH 15 - BRAZIL'S FOREIGN TRADE WITH PARAGUAY AND URUGUAY (US$-FOB)
SOURCE: MDIC/SECEX.

Since the 1980s, Paraguay's imports have come from MERCOSUR member countries, accounting for 47.4% in 1980, with a period of decline from 1985 onwards (30.09%), and then an upturn between 1991 and 1996 (52.85%). In 2002, this percentage reached its peak (55.2%) and then began to decline, representing 35.41% in 2006. The country that most increased its exports to Paraguay, taking the place of MERCOSUR, was China, which went from 11.44% in 2000 to 25% in 2006, also contributing to the decline in exports from the United States and Canada and Europe, which in 2006 accounted for 5.99% and 8.76% respectively (FARIA; COUTINHO, 2009).

Uruguay also has a predominance of imports from MERCOSUR countries, which have grown considerably, from 33.06% in 1980 to 57.53% in 2007, with only one period of significant decrease between 1981 (37.69%) and 1985 (27.98%). During this period of decline, those who increased their exports to Uruguay the most were the Arab and Islamic countries, growing their share from 1.06% to 16.77%, Africa, from 11.77% to 15.58%, and the Central American countries, from 1.81% to a peak of 9.29% in 1983 (FARIA; COUTINHO, 2009).

In the case of the relationship between Brazil and Uruguay, the most common products traded are cattle, soybeans and food products. According to the researchers Haesbaert and Barbara (2001), Brazilian entrepreneurs often have businesses on both sides of the border.

Despite its structural and economic problems, MERCOSUR represented the first South American integration process, and also a Latin American one, to achieve concrete results and open up regional alternatives for a better international integration of the Southern Cone countries within the framework of an emerging world order (VIZENTINI, 2007).

Argentina, for example, has become the preferred destination for Brazilian companies at the start of their internationalization strategies since 2001-2002. The total volume of investments made by Brazilian companies in Argentina in the period 1997-2008 was approximately US$ 9 billion, according to the Brazilian Embassy in Buenos Aires in contact with companies and through official agencies. These figures include new projects, mergers and acquisitions, reinvestments and expansions (MDIC, 2010).

From the point of view of regional integration, there are various factors that act to give consistency to or weaken economic policy measures aimed at reserving the domestic market, or even expanding its dimensions beyond the territorial borders of the nation state through economic integration. One of these factors is the political and economic power held by certain fractions of capital to impose their rules of articulation and integration with the world market (EGLER, 2001).

For a geographer who is attentive to the game of scale, this means that economic, geopolitical and cultural factors play out equally on a global and regional scale. Because the great changes in scale in the past, the broadening of the geographical and economic horizons of states are always accompanied by a formalized *decoupage of* the world. It is necessary to know where the sphere of influence lies (FOUCHER, 2011).

CHAPTER 6

6. INTERNATIONAL TRADE IN BORDER MUNICIPALITIES

No nation has yet been ruined by trade.
Benjamin Franklin

The relationship between foreign trade and the border municipalities shows that the spatial interactions of the border can be understood on different scales, as they are part of a broad and complex commercial network linked through the flows emitted by each municipality. In this sense, this chapter aims to demonstrate the intensification of foreign trade flows in border municipalities after the country's economic opening up and the construction of economic blocs, showing the main results of the paradigm shift towards development. Economic relations on different scales (local, national and international) indicate a change in the spirit of the times, where until a few years ago the border had more closed geopolitical characteristics and the data in this chapter shows that the development paradigm (from an economic point of view) is present in several Brazilian municipalities.

The municipal statistical analysis covers the period from 1999 to 2013, although the integration of the economic blocs took place before that year, but unfortunately the data by municipality is only available since then. The chapter describes the game of scale that exists on the border, starting with the flows generated on a national scale and moving on to an analysis by border arc, and then by municipality, highlighting the evolution of trade flows and the main trading partners.

6.1 Evolution of Brazil's foreign trade flows

Throughout history, international trade has played an important role in relations between countries and has helped to reduce spatial and geographical barriers. Since the 1990s, in the context of economic integration, the result of a privileged relationship between countries and between economic blocs, customs and non-customs barriers have been minimized because, as economic integration deepens, the flow of goods, capital and people accelerates.

Driven by globalization, world trade and economic flows have increased significantly in several countries. In Brazil in particular, trade flows (generated by the import or export of products and/or services) have also grown significantly since the opening up of trade, but it is still considered a small opening.

Foreign trade flows generate what Trevisan (2012) defines as Newton's third law, applied to foreign trade, i.e. every import corresponds to an export, of the same intensity, but in the opposite direction. In this way, outflows represent exports for a country and imports for the receiving country.

Brazilian exports grew in absolute terms between 1999 and 2013, with the exception of 2009, due to the global crisis. Graph 16 shows the evolution of Brazil's exports and imports between 1999 and 2013. As a curiosity, in 1999 Brazil exported 48 billion FOB/US$ and in 2013 it exported 242 billion. The highest growth in Brazilian exports was 32% between 2009/2010 and 2003/2004. However, during the global crises of 2008/2009 and 2011/2012 there was a drop in exports.

In terms of imports, Brazil imported 49 billion in 1999 and 239 billion FOB/US$ in 2013. The highest rate was between 2007/2008, with growth of 43%, with a relative drop during the world crisis of 2008/2009. In the period analyzed, the volume of growth in imports varied more than the volume of exports. The reasons for this may be related to the economic policies applied, restrictions on certain products, restrictions in customs codes, exchange rate policy and the pressure exerted by entrepreneurs when they want to "protect" their consumer market.

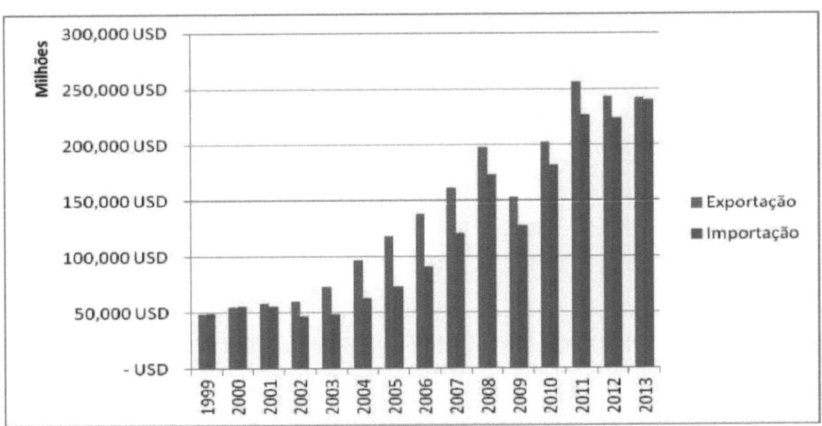

GRAPH 16 - EVOLUTION OF BRAZIL'S EXPORTS AND IMPORTS FROM 1999 TO 2013
(US$-FOB)

Source: MDIC/SECEX

International trade brings benefits to the country. These include, for example, balancing the balance of payments, updating technology, increasing exports and reducing dependence on the domestic market. The balance of payments is a tool used to record all the imports Brazil has made from other countries, and all the exports Brazil has made to other countries. It also records the loans made, the capital of foreign firms that open branches in the country, the capital of foreign firms that leave Brazil, among others.

The balance of payments is often used to indicate the "health" of a country's foreign trade. Exports are recorded with a (+) sign and imports with a (-) sign. When the balance of payments is in deficit, countries become more exposed to global crises, so product diversification is desirable because, in addition to balancing the balance of payments, countries with high technology are able to add more value to their goods.

Faced with the growing changes influencing the dynamics of international trade, governments and companies have been looking for strategies to guarantee gains in competitiveness, access to new markets, a reduction in operating risks and new sources of funding.

For the entrepreneur, foreign trade can bring benefits such as better use of the company's idle capacity,
market diversification, increased productivity, product adaptation, less dependence on the domestic market, taking advantage of export incentives and, in some cases, a reduction in the tax burden.

Given this panorama, the internationalization of companies plays a crucial role, especially for emerging economies that are formulating policies for sustainable economic growth. The internationalization of production occurs when residents of one country gain access to goods and services originating in another (MDIC, 2009).

Between 1999 and 2013, Brazilian exports and imports grew, as already discussed. In addition to the intensification of these flows, the number of exporting companies increased by 18.74% between 2001 and 2013 and importing companies grew by 38%. The average number of exporting companies in Brazil is 20,604 and the average number of importing companies is 3,970.

If you compare this to the total number of companies in Brazil, which is around 16 million, the number of "internationalized" companies is not even 1% (both in terms of imports and exports). This shows that Brazil is still a very closed country to the foreign economy. Graphs 17 and 18 show the evolution of the number of exporting and importing companies in Brazil.

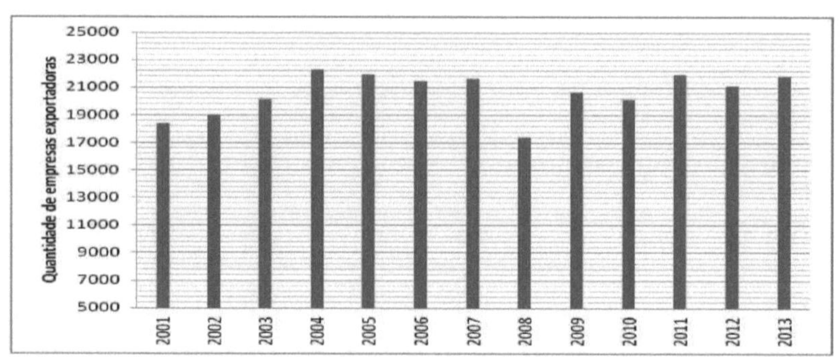

GRAPH 17 - NUMBER OF EXPORTING COMPANIES IN BRAZIL BETWEEN 2001 AND 2013
SOURCE: MDIC/SECEX

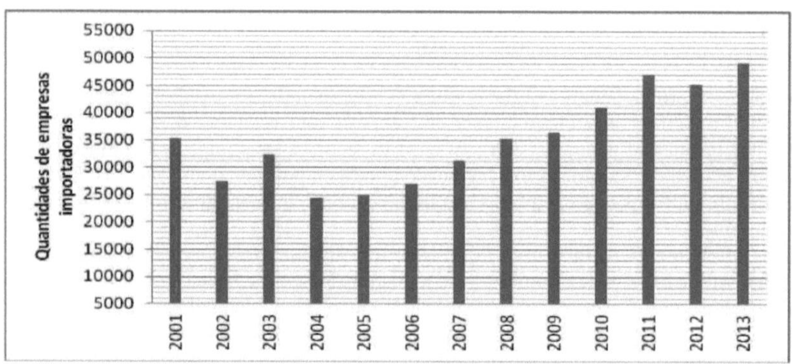

GRAPH 18 - NUMBER OF IMPORTING COMPANIES IN
BRAZIL BETWEEN 2001 AND 2013
SOURCE: MDIC/SECEX

One of the consequences of foreign trade is that over time there are changes in the profile of foreign investments and in the nature of foreign flows of capital (financial and productive), technology and goods in general, accelerated by the opening up of trade, which depends on a combination of domestic policies and the process of economic liberation (COSTA, 1999).

According to Catela (2009), various empirical studies point to the greater efficiency of firms that export compared to those that focus exclusively on the domestic market. The first hypothesis is that the most productive firms are self-selected as exporters. The reason for this is that there are additional costs when selling products on foreign markets, including transportation costs, costs associated with establishing distribution channels and production costs for modifying goods to suit the tastes of foreign customers.

The second hypothesis is that exporting offers the opportunity to take advantage of *learning by exporting*, which plays a fundamental role in improving productivity. Firms that participate in international trade are exposed to more intense competition than those that only sell on the domestic market, so they must improve their results in order to remain on the foreign market. The results show that strongly exporting firms exhibit productivity 2.3 times higher than that of other exporters, while firms oriented towards the domestic market achieve half the productivity of potential exporting firms (CATELA, 2009).

Another characteristic concerns the quality of exports. Exporters tend to export more technologically-intensive products than the others and manage to get these products into more demanding markets such as the USA, Canada and the European Union (FERREIRA, 2009).

With the aim of increasing Brazilian exports - both industrial and services - the Federal Government has directed actions such as Incentive to Export, Foreign Trade Meetings (ENCOMEX), Exporter's Portal, Exporter's Showcase, First Export Project, Foreign Trade Information Centre (CICEX), IMPE Portal (Internationalization of Micro and Small Companies), ApexBrasil Portal, Brazilian Network of International Business Centres, International Trade *Centre (*ITC). The flow of exports is expected to increase significantly, but the inclusion of more companies to export is still very small.

The Ministry of Development, Industry and Foreign Trade is responsible for organizing and regulating foreign trade in Brazil. In addition to formulating proposals for foreign trade policies and programs and establishing rules for their implementation, the MDIC proposes measures and guidelines that articulate the customs instrument and trade relations with other countries. The government also applies a tax policy to encourage exports by reducing taxes on products. According to Sebrae-SP (2013), exporting industries receive the following incentives:

a) exported products are not subject to the Industrialized Products Tax (IPI);

b) The Tax on the Circulation of Goods and Services (ICMS) is not levied on exports of industrialized products, semi-manufactured products, manufactured products, primary products or the provision of services;

c) in determining the basis for calculating the Social Security Financing Contribution (COFINS), revenue from exports is excluded;

d) revenues from exports are also exempt from contributions to the Social Integration Program (PIS) and the Public Servant Equity Formation Program (PASEP);

e) The Tax on Financial Transactions (IOF) applied to foreign exchange transactions linked to the export of goods and services has a zero rate.

In general, the flows generated by foreign trade are supported by structuring factors that support them and intensify trade. These include exchange rate policy, which is important for boosting the international economic system.

Exchange rate policy has a direct impact on the trade balance and, consequently, on the balance of payments. With regard to the exchange rate regime, Brazil has adopted a floating exchange rate, i.e. the price of the dollar varies according to market fluctuations. Under this regime, the exchange rate is determined by the relationship between supply and demand. The value of the foreign currency changes when the rate rises due to increased demand for dollars, and the national currency devalues.

According to the institutional model, the Central Bank of Brazil is responsible for defining the exchange rate regime, its targets and

management, i.e. exchange rate policy. According to the Central Bank of Brazil (BACEN, 2013), the Brazilian exchange rate system has been floating since February 1999, with the BCB intervening only occasionally. Transactions on the foreign exchange market are carried out by banks, brokers and travel agencies authorized by the BCB, the latter two only operating with paper money and traveller's cheques. Banks are allowed to operate in the foreign exchange futures market, but within limits set by the BCB and these operations must be settled within 360 days.

6.2 **FOREIGN TRADE FLOWS GENERATED ON THE BORDER STRIP**

The relationship between foreign trade and the national territory becomes explicit when studying border regions because spatial interactions on the border can be understood on different scales (local, national and international), in a broad and complex set of flows of people, goods, capital and information over geographical space.

Analyzing the export flows of the border municipalities, the flow of goods and their exponential growth over time is clear. In 1999 the flow of exports was 1.4 billion dollars, in 2013 it was 15 million, or 10 times more than in the first year studied. Looking at another period, it can be seen that between 1999 and 2006 the flow of exports increased by 150% and between 2006 and 2013 it increased 5 times. Among the years with the best performance were 2008 and 2013.

The scale of export flows shows us the greater dynamism and interaction of border towns in the local economy and their links with the international trade network. Thus, the growth in export flows in FOB value brings a new dynamism to the border strip.

The increase in the flow in quantitative terms is the result of several factors, but the main ones are the exchange rate policy adopted and the expansion of Brazil's trade relations. The relative percentage between exports from border municipalities and total exports from Brazil in 1999 was 3.05%, in 2013 the ratio was 6.5%; this means that export flows were more important in 2013 in percentage terms on a national scale, which means that in 14 years the relative importance of flows has increased, making it a more dynamic region and more interconnected with other countries. Graph 19 shows the evolution of the flow of exports from the municipalities along the border.

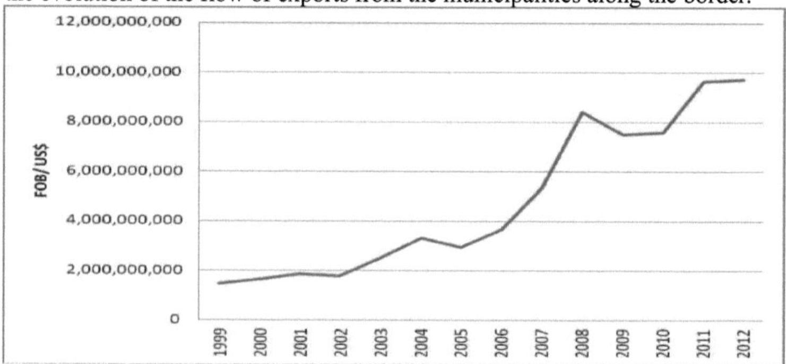

GRAPH 19 - EVOLUTION OF EXPORT FLOWS FROM BORDER STRIP MUNICIPALITIES (FOB-US$)
SOURCE: MDIC/SECEX

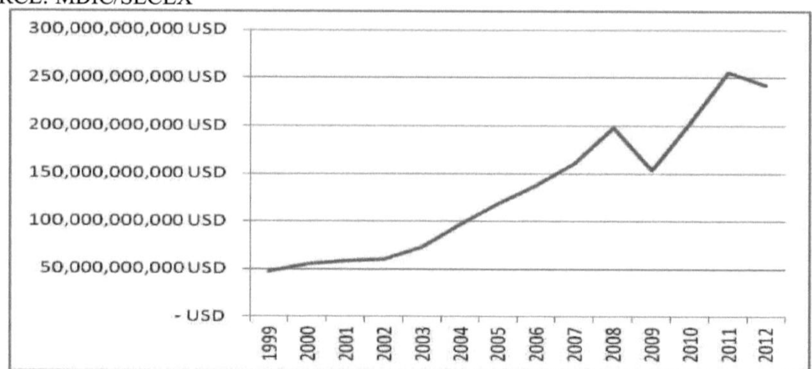

Bringing Graph 16 back into the picture, which deals with Brazil's export flow, we can see that the dynamics on a national scale and on a regional scale (border strip) had a similar result. The first characteristic is that both showed an upward growth in the flow of exports. The second characteristic is that on both scales there was a drop in the flow in 2008/2009, as a result of the global crisis.

Continuing along the line of argument, it can be seen that the relative importance of the flow generated on the border strip in relation to Brazil's total export flow between 1999 and 2013 can be seen in Graph 20.

The relative percentage between exports from border municipalities and total exports from Brazil in 1999 was 3.05%, in 2013 the ratio is 6.5%; this means that the export flows generated in border municipalities have increased proportionally more than the volume initially generated in 1999, i.e. the volume of exports at the border has increased more than at the national level in proportional terms.

Another relevant fact is that during the period of the global crisis in 2008, the proportional volume of exports from border municipalities was higher in this period, which means that the impact of the crisis was greater on a national scale than on a regional scale. In 2005, the relative percentage was 2.48, the lowest achieved in the period analyzed.

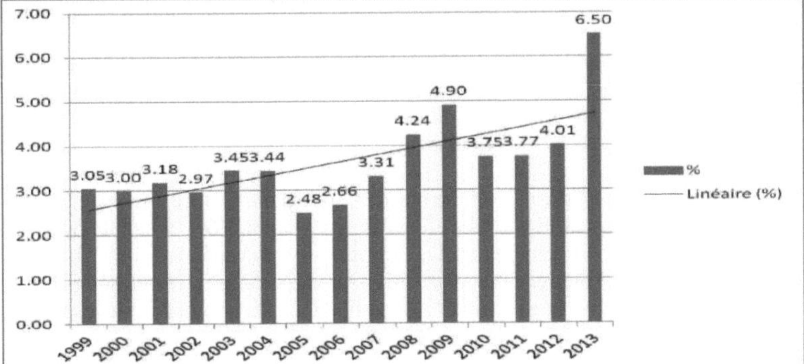

GRAPH 20 - RELATIVE PERCENTAGE OF EXPORTS FROM BORDER MUNICIPALITIES IN RELATION TO TOTAL EXPORTS FROM BRAZIL
SOURCE: MDIC/SECEX

The number of countries to which the border municipalities exported during the period analyzed can be seen in Table 09. It can be seen that the number did not vary and the average is 160 countries on various continents: Europe, America, Africa, Asia and Oceania.

It is not possible to know from this data alone what the reasons were for opening up, maintaining or losing these trade flows generated because this depends on a number of factors, such as knowledge of the consumer market, the ability of Brazilian companies to adapt, the commercial needs of the client, exchange rate issues, commercial intelligence issues, but it is possible to say that export flows are multidirectional in all the years analyzed.

TABLE 09 - NUMBER OF COUNTRIES TO WHICH THE MUNICIPALITIES ON THE BORDER STRIP EXPORTED BETWEEN 1999 AND 2013

Year	Number of countries
1999	161
2000	124
2001	121
2002	153
2003	153
2004	170
2005	172
2006	172
2007	182
2008	178
2009	175
2010	175
2011	175
2012	171
2013	170

Source: AliceWeb.

Analyzing export flows by country, in order of importance, according to FOB-US$ value, in 1999 the main trading partners of the border municipalities were Paraguay (19.74%), Argentina (13.10%), Spain (10.88%) and the United States (10.79%), which together accounted for 54.51% of the total generated. Flows were directed to 161 countries, but more than half of the flow in quantitative value is concentrated in just 4 countries, 32.84% of which was directed to neighboring countries and members of MERCOSUR.

In 2006, the main trading partners of the border municipalities were China (16.17%), the United States (16.59%), Iraq (11.90%) and Russia (10.85%), together amounting to 1.2 billion dollars. The export flow was generated for 172 countries, with the Southern Cone countries losing relative importance compared to 1999. The flow generated for South American countries this year was 611 million dollars.

In 2013, the main trading partners of the border municipalities were Panama (27.04%), China (23.36%) and the Netherlands (23.01%), as well as Hong Kong (4.86%), which together accounted for 8.2 billion dollars of the total generated. The export flow was generated for 170 countries, with the flow generated for South American countries this year amounting to 1.4 billion dollars. In other words, it can be said that there has been economic dynamism in some border municipalities as a result of their inclusion in the international trade circuit.

Geographical proximity and tariff agreements may be relevant in analyzing the result. Among South American countries, in order of importance in terms of total flow, Paraguay and Argentina took turns as the main trading partners between 1999 and 2013. From 1999 to 2002, 2006 to 2008 and 2011 to 2013, Paraguay was the leader in the export flow *ranking*. Argentina was the leader from 2003 to 2005 and from 2009 to 2010. Exports from border municipalities to MERCOSUR member states can be seen in Graph 21, with growth rising between 2002 and 2007, then falling in the 2008 crisis, with growth resuming from 2009 onwards.

Geographical proximity and tariff agreements are relevant in the analysis of the result, and it can be seen that there is an influence from the conduct of foreign policy, which in some years interferes more than geographical proximity. Among the Southern Cone countries, in order of importance of the total flow, Paraguay and Argentina took turns as the main trading partners between 1999 and 2013, but flows to the Southern Cone have been losing relative importance.

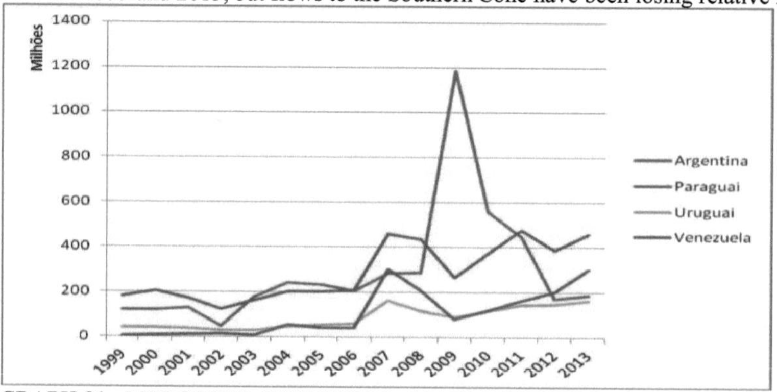

GRAPH 21 - MERCOSUR COUNTRIES - MAIN PARTNERS - (FOB-US$)
SOURCE: MDIC/SECEX

Between 1999 and 2006 the flow of exports to South American countries increased by 55% of the total value. Analysis of the export flow between 1999 and 2006 shows that the highest growth rate in exports from border municipalities to South American countries was from Peru, Colombia and Chile. Between 2006 and 2013, the flow of exports to South American countries grew by 139%. Table 10 shows the flow of exports at the border by South American country.

TABLE 10 - EXPORT FLOW BY SOUTH AMERICAN COUNTRY IN 1999, 2006 AND 2013

	1999	2006	2013
Argentina	121,084,447 USD	206,384,147 USD	188,443,191 USD
Bolivia	28,849,37 USD	30,770,098 USD	122,948,365 USD
Chile	9,727,293 USD	38,875,076 USD	83,592,167 USD
Colombia	319,083 USD	5,797,349 USD	73,001,525 USD
Ecuador	0 USD	7,612,900 USD	23,525,927 USD
Guyana	181,382 USD	202,766 USD	2,750,608 USD
French Guiana	74,598 USD	28,325 USD	146,560 USD
Paraguay	182,506,773 USD	209,875,057 USD	459,015,609 USD
Peru	1,484,634 USD	9,687,559 USD	45,490,519 USD

Suriname	0 USD	197,763 USD	701,450 USD
Uruguay	43,053,097USD	60,668,950 USD	164,293,175 USD
Venezuela	4,883,669 USD	41,031,139 USD	301,349,724 USD
Total	392164352	611131129	1,465,258,820 USD
Source: AliceWeb, 1999, 2006, 2013.			

Although the flow of exports from border municipalities to South America increased between 2006 and 2013, the relative importance of South America has been decreasing since 2009, according to Graph 22.

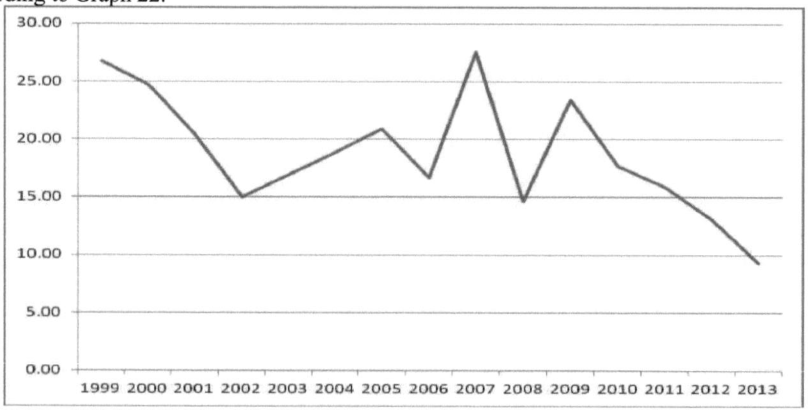

GRAPH 22 - RELATIVE SHARE OF EXPORTS FROM BORDER MUNICIPALITIES TO SOUTH AMERICAN COUNTRIES - (FOB-US$)
Source: AliceWeb - author's elaboration.

On a regional scale, exports from the Central Arc have been on the rise since 1999. In 1999, its municipalities exported approximately 64 million dollars; in 2013 they exported approximately 2.5 billion. There was a drop between 2009 and 2010, probably due to the global crisis (Graph 23).

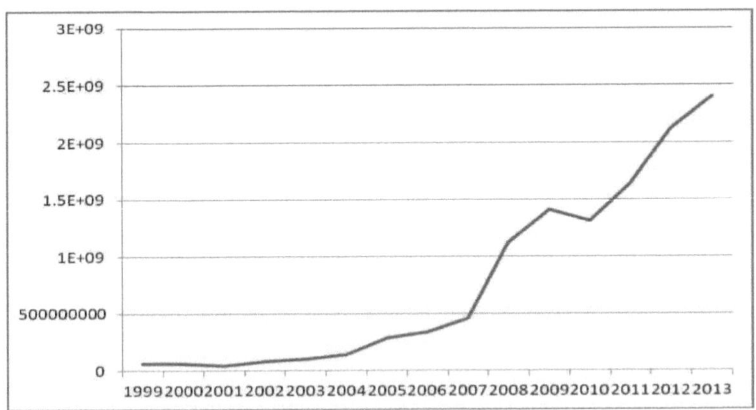

GRAPH 23 - TOTAL EXPORTS OF THE CENTRAL ARC BORDER MUNICIPALITIES
BETWEEN 1999 AND 2013 - (FOB- US$)
Source: AliceWeb. Prepared by the author.

In Arco Norte, exports were on the rise from 1999 to 2008. In 1999 exports were 205 million dollars and in 2008 they were 684 million dollars. In 2009 there was a 44% drop, down to 412 million. Between 2009 and 2012 it grew again, reaching 921 million in 2009, falling again from 2010 onwards (Graph 24).

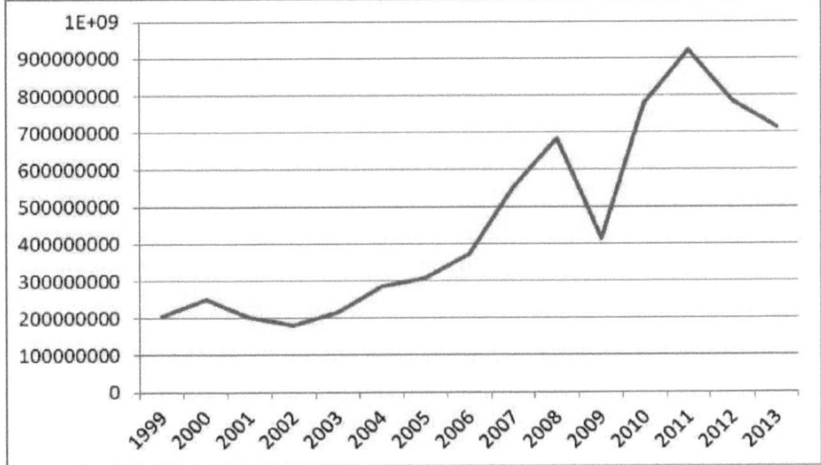

GRAPH 24 - TOTAL EXPORTS OF NORTHERN ARC BORDER MUNICIPALITIES
BETWEEN 1999 AND 2013 - (FOB-US$)
Source: AliceWeb - author's elaboration.

The Southern Arc, the most dynamic, with the largest export flow, exported 1.1 billion dollars in 1999. In comparison, in the same year, this is practically the sum of the exports of the Northern Arc and the Central Arc. In 2013, the Southern Arc exported 12.6 billion dollars (Graph 25).

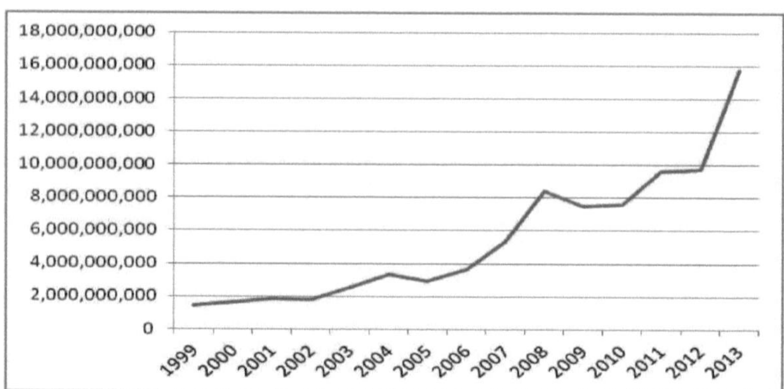

GRAPH 25 - TOTAL EXPORTS FROM SOUTHERN ARC BORDER MUNICIPALITIES BETWEEN 1999 AND 2013 - (FOB-US$)
SOURCE: AliceWeb. Prepared by the author.

The Southern Arc is certainly the most commercially dynamic, as it is home to the majority of exporting companies, the majority of the population and the majority of border towns. The Southern Arc experienced a relative drop in exports between 2008 and 2009, probably due to the global crisis. Between 2012 and 2013 there was a growth rate of 84%.

In 1999, the Central Arc's main trading partners were Bolivia, with 16.9 million dollars, followed by Argentina, with 6.2 million dollars and Germany, with 6.1 million dollars. The main partners in South America were Bolivia (16.9 million), Argentina (6.2 million) and Uruguay (365 thousand). Exports from the Central Arc accounted for only 5.40% of total exports from border municipalities, while exports from South American countries accounted for 6.02% of the total.

The Northern Arc's main trading partners are: Canada, with 46.6 million, Belgium, with 3.4 million and the United States, with 2.5 million. In total, the Northern Arc exported $187 million. Among the Southern Cone countries, the main trading partners were Venezuela (2.3 million), Bolivia (799 thousand) and Argentina (364 thousand). In the Northern Arc's total, the Southern Cone countries represent 0.88% of export destinations with a turnover of 3.4 billion.

In the Southern Arc, the main partners in South America were: Argentina with 53.9 million, Paraguay with 182.5 million and Spain with 92 million. In total, the flow of exports from the Southern Arc was 829 million, which represents 77.16% of the total generated in the municipalities. Among the Southern Cone partners, in order of importance are Paraguay, Argentina and Uruguay (42.6 million). The total flow to the Southern Cone countries is 363 billion, representing 93.09% of the total.

In 2006, the main trading partners in the Central Arc were China with 285 million, the United States with 45.7 million and Italy with 41.7 million. In total there were 239 billion in exports, with the Central Arc accounting for 10.19% of total exports. The Southern Cone countries alone were Argentina with 5.3 million, Bolivia with 13.22 million and Venezuela with 1.9 million.

The main partners in the Northern Arc were the United States with 128 million, Belgium with 56.9 million and Italy with 40.9 million. In total, 341 million of the total export flow came from the Northern Arc, which represents 14.54%. Its main trading partners in the Southern Cone were Bolivia with 9.9 million, Venezuela with 8.5 million and Peru with 234 thousand.

In the Southern Arc, the main partners were Argentina with 200 million, China with 302 million and Iran with 253.7 million. The total flow was 1.7 billion and this represents 75.27% of the total export flow. Among the Southern Cone countries, Paraguay with 209 million, Argentina with 200 million and Uruguay with 59 million. A total estimate of 570 million for the Southern Cone.

In 2013, the Central Arc's main trading partners were Venezuela with 244.5 million, Hong Kong with 207 million and Switzerland with 181 million. The total export flow was approximately 1.4 billion, which represents 14.34% of the border strip's total export flow. Among the Southern Cone countries, the main partners were Bolivia with 75.4 million, Venezuela with 244 million and Colombia with 45.5 million, out of a total of 378 million from the Southern Cone countries.

In the Northern Arc, the main partners were China with 126 million, Switzerland with 101 million and Hong Kong with 66 million. The Northern Arc represents 5.2% of the total export flow. The total export flow was 582 million dollars. Among the Southern Cone countries, the most representative were Bolivia with 12.1 million, Peru with 3.2 million and Venezuela with 2.8 million out of a total of 20.2 million.

The main trading partners in the Southern Arc are Panama, with 2.8 billion dollars, the Netherlands, with 2.2 billion dollars and China, with 2.1 billion dollars, totaling 9.3 billion dollars, which represents 82.54% of the total flow of exports from the border municipalities. Among the Southern Cone countries are Paraguay with 458 million, Argentina with 185 million and Uruguay with 16.2 million, totaling close to 1 billion dollars.

In the reverse flow of goods, import flows on the border strip show an upward trend, with 2008 being the highlight, as can be seen in Graph 8.11. Between 1999 and 2013 the flow of imports grew exponentially. Between 1999 and 2006 the flow increased by 49% and between 2006 and 2013 it increased 3.5 times. Graph 26 shows the evolution of the flow of imports from the municipalities on the border strip.

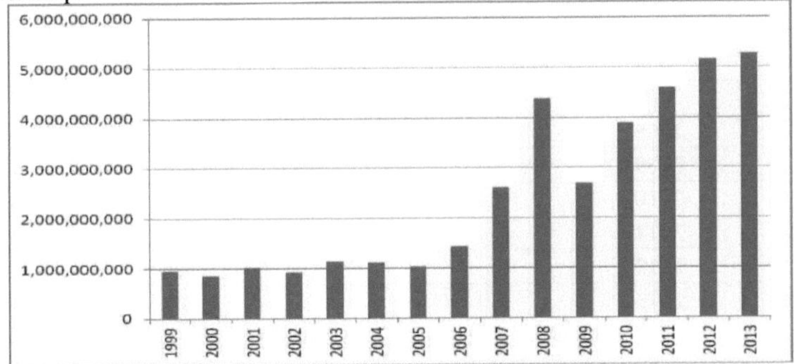

GRAPH 26 - EVOLUTION OF THE FLOW OF IMPORTS FROM BORDER AREA MUNICIPALITIES - (FOB-US$)
SOURCE: AliceWeb. Prepared by the author.

The growth in import flows in FOB value occurred mainly after 2006. The relative percentage of the flow generated on the border strip in relation to Brazil's total import flow between
1999 and 2013 can be seen in Graph 27. The relative percentage between imports from border municipalities and total imports from Brazil in 1999 was 1.95%, while in 2013 the ratio was 2.20%, meaning that border municipalities are not the main importing municipalities in Brazil. The relative percentage of imports is lower than the relative percentage of exports.

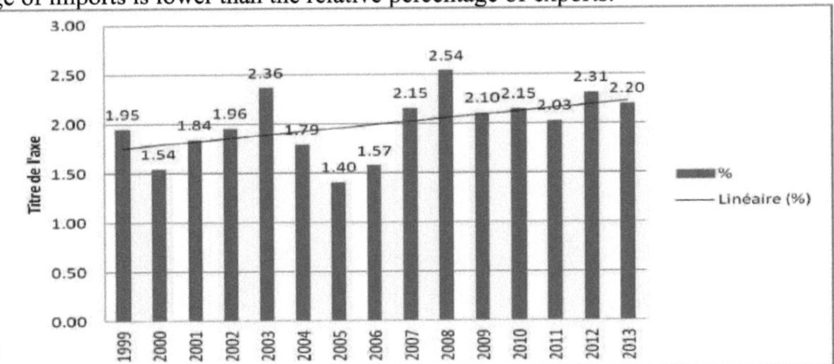

GRAPH 27 - RELATIVE PERCENTAGE OF IMPORTS FROM BORDER MUNICIPALITIES IN

RELATION TO TOTAL IMPORTS FROM BRAZIL

SOURCE: MDIC/SECEX. Own elaboration

Table 11 shows the number of countries from which border municipalities imported during the period under analysis. It can be seen that the number of countries included in the import flow is smaller than the number included in exports.

TABLE 11 - NUMBER OF COUNTRIES IN WHICH THERE WERE IMPORTS TO BORDER MUNICIPALITIES

Year	Number of countries
1999	69
2000	50
2001	72
2002	75
2003	75
2004	76
2005	75
2006	84
2007	85
2008	95
2009	94
2010	101
2011	103
2012	105
2013	99

Source: AliceWeb. Own elaboration

On a regional scale, it can be seen that in 1999 the Central Arc's main trading partners for imports were Cyprus, with 1.7 million, the United States, with 1.3 million and Germany, with 926 thousand, totaling 6.7 million, which represents 0.80% of the total imported by the border municipalities. The main partners in South America were Bolivia (757,000), Argentina (838,000) and Uruguay (45,700), totaling 1.6 million. Graph 28 shows the evolution of imports in the central arc between 1999 and 2013.

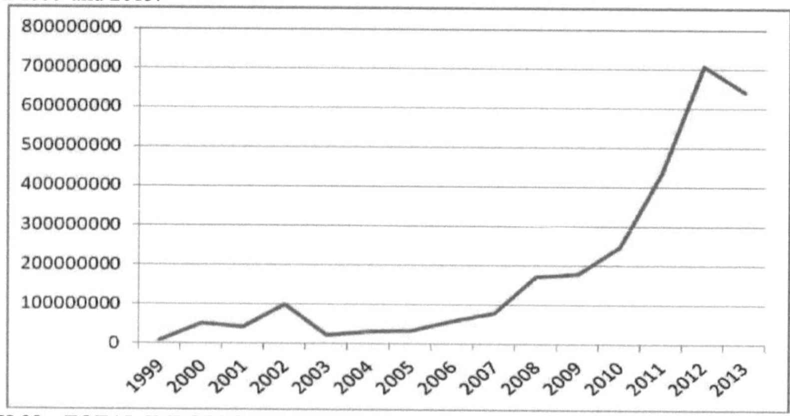

GRAPH 28 - TOTAL IMPORTS BY CENTRAL ARC BORDER MUNICIPALITIES BETWEEN 1999 AND 2013 - (FOB- US$)

SOURCE: MDIC/SECEX

In 2006, the main trading partners in the Central Arc were Israel, with 10 million, China, with 8 million and the United States, with 5.4 million, totaling 48 million of the import flow. The Southern Cone countries alone were Argentina with 3.9 million, Bolivia with 515 thousand and Uruguay with 75 thousand.

In 2013, the Central Arc's main trading partners were: China, with 252 million, the United States, with 117 million and South Korea, with 26 million, for a total of 514 million, representing 12.93% of the border strip's total export flow. Among the Southern Cone countries, the main partners were: Argentina, with 9.1 million, Colombia, with 9.4 million and Chile, with 2.1 million. In total there were 23 million from South American countries.

In the Northern Arc, in 1999 the main trading partners were Venezuela, with 1.9 million, Japan, with 1.4 million and the United Kingdom, with 1.07 million, totaling 10.5 million, which represents 1.25% of total imports on the border strip. Among the South American countries, the main trading partners were Venezuela (1.9 million), Bolivia (13 thousand) and Argentina (338 thousand). The flow from the Southern Cone countries on the border strip in 1999 was 472 million dollars. Graph 29 shows the flow of imports in the northern arc.

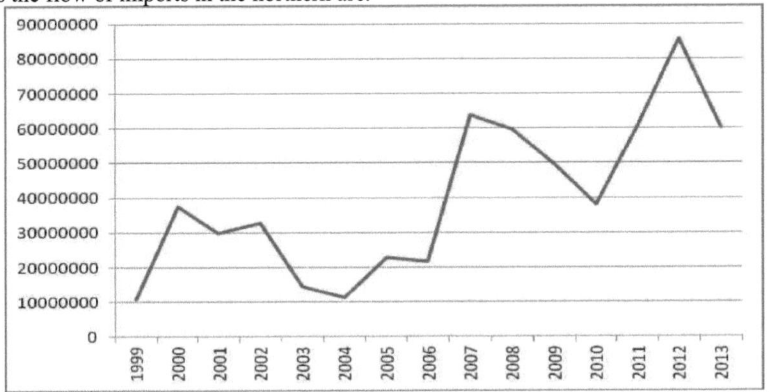

GRAPH 29 - TOTAL IMPORTS BY NORTHERN ARC BORDER MUNICIPALITIES BETWEEN 1999 AND 2013 - (FOB-US$)
Source: MDIC. Prepared by the author.

In 2006, the main partners were Japan and Germany, with 2 million and the United States, with 6 million, totaling 3 million in imports. South America's main trading partners were Venezuela with 382 million, Argentina with 14.3 million and Uruguay with 636 million.

In 2013, the main partners were China with 16.5 million, the United States with 16 million and Germany with 6.8 million, for a total of 55 million. Among the Southern Cone countries, the most representative were Venezuela with 627,000, Peru with 189,000 and Chile with 239,000, totaling 1.5 million.

In 1999, South America's main partners in the Southern Arc were: Argentina, with 356 million, the United States, with 226 million and Uruguay, with 56 million, totaling 825 million, which represents 97.95% of the border strip's import flow. Among the Southern Cone partners, in order of importance: Argentina with 356 million, Uruguay with 56 million and Paraguay with 35 million. The total flow of imports from the Southern Cone countries was 472 million. Graph 30 shows the evolution of import flows in the southern arc.

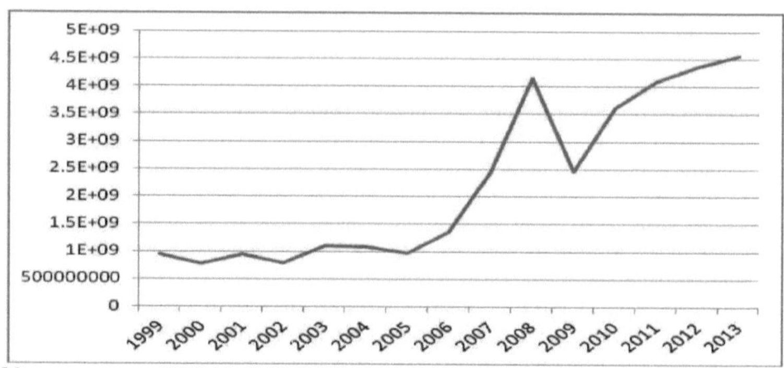

GRAPH 30-   TOTAL IMPORTS BY MUNICIPALITIES
SOUTHERN ARC BORDERS BETWEEN 1999 AND 2013 - (FOB-US$)
Source: MDIC. Prepared by the author.

In 2006, the main partners were Argentina with 469 million, Paraguay with 13 million and Morocco with 106 million, totaling 1 billion in trade. Among the South American countries, Argentina with 469 million, Paraguay with 113 million and Bolivia with 38 million. The flow from South American countries was 705 million.

In 2013, the main trading partners were Argentina with 1 billion, the United States with 466 million and Mexico with 382 million, for a total of 3 billion. Among the South American countries are Argentina, with 1 billion, Chile, with 88 million and Paraguay, with 290 million, totaling close to 1.5 billion dollars.

Foreign trade flows are influenced by different orders and factors, which can be of a political, economic, tax, exchange rate and social nature. Each factor influences a specific region in a particular way. It can be seen that in the arcs analysed, international trade is not just a tool, but an agent that intensifies the flow of goods in the context of the era of globalization because the expansion of capitalism has resulted in its growth, intensified by relations between the most varied countries.

6.3 FOREIGN TRADE NETWORKS IN BORDER CITIES

Geography is directly related to commercial development and the urban dynamics of the cities that absorb commercial and international flows. For author Saskia Sassen (2007), a "mega-regional" scale can help connect the winners and laggards in a mega-region that becomes cities and areas, both globalized and local. This connection can also be converted into cross-border intercity networks by strengthening connections, since different circuits contain different groups of countries and cities. Seen in this way, the global economy becomes concrete and specific, with a well-defined geography. These circuits are multidirectional and criss-cross the world, feeding intercity geography as strategic nodes.

The competitiveness of cities depends not only on the country's macroeconomic conditions, but also on geographical and environmental characteristics. The great modifying agent is the companies grouped together in cities and regions that constitute the engine of the new economic system that is prevailing. Issues of power and international politics within strategic networks have been highlighted by literature based on international political economy (SASSEN, 2007).

Urban networks have particularities inherent to their socio-spatial formation, and can be associated with more complex networks. Both the processes triggered on a local/regional scale and those produced on other broader spatial scales have repercussions on the formation and development of cities and the definition of functions in the network. Many of the experiences reported in international literature treat specialized territorial economies as a type of product or production chain in the regional geo-economic fabric. In fact, the mechanism of international specialization has consequences for regional dynamics and selects the areas where skills and knowledge will be accumulated, which can generate divergent technological trajectories, with different potential for growth and innovation (BREITBACH, 2005).

To understand the relationship between cities and the global economy, it is useful to specify the multiple global circuits in which cities are connecting borders. Particular networks connect particular groups of cities and this makes it possible to recover details about the various roles of the city in the global economy (SASSEN, 2007).

Considering the connection that exists between the international economy and cities, it can be seen that in the case of Brazilian border municipalities, not all of them are connected through the flow of foreign trade to another country. And the border towns that are connected through the flow of foreign trade to another country are connected in a very particular way. Tables 12 and 13 show the number of border towns that exported[49] and imported, respectively, between 1999 and 2013, by border arc. The table does not distinguish between municipalities that exported and imported at the same time.

TABLE 12 - NUMBER OF BORDER TOWNS THAT GENERATED EXPORT FLOWS BETWEEN 1999 AND 2013, ACCORDING TO BORDER AREA

Year	Central	North	South
1999	17	14	152
2000	19	20	156
2001	22	20	177
2002	24	19	174
2003	25	17	176
2004	21	20	176
2005	22	18	170
2006	19	19	167
2007	18	21	176
2008	21	21	168
2009	21	15	165
2010	23	15	152
2011	22	15	164
2012	22	15	163
2013	22	14	168

Source: AliceWeb. Prepared by propna

TABLE 13 - NUMBER OF BORDER CITIES PARTICIPATING IN THE IMPORT FLOW BETWEEN 1999 AND 2013 BY BORDER ARC

	Central Arch	Northern Arc	Southern Arc
1999	12	8	128
2000	11	10	133
2001	11	7	132
2002	111	8	131
2003	10	11	134
2004	11	10	138
2005	11	9	137
2006	12	9	137
2007	13	10	140
2008	16	11	153

---

[49] Exporting municipalities are considered to be all those that have had an export process legally registered and have their tax domicile in that municipality. Importing municipalities are all those that have had an import process legally registered and have their tax domicile in that municipality.

2009	14	11	147
2010	17	13	156
2011	15	13	165
2012	13	13	175
2013	14	11	178

Source: AliceWeb. Own elaboration

The municipalities that generate foreign trade flows are part of a system of cities linked to international trade. Between 1999 and 2013, an average of 205 border towns were included in the foreign trade network[50] , i.e. almost half of the border towns. The networks formed by the border towns in each arc/state can be seen in Figures 16 to 20, which show the relationship between the border towns and the various countries.

The networks formed by border cities are particular because they select cities according to their geographical, urban, business and logistical conditions to be part of the international trade network practiced on the border. Each border arc forms a foreign trade network linked to different countries. The Southern Arc is the one with the highest concentration of municipalities participating in the foreign trade network. This is because it is home to the largest number of companies, the most populous states and the ones with the most municipalities on the border. The number of exporting municipalities in Arco Sul is, on average, 10 times greater than in Arco Norte and 9 times greater than in Arco Central. In terms of import flows, the number of importing municipalities in the Southern Arc is, on average, 14 times higher than in the Northern Arc and 9.8 times higher than in the Central Arc.

The trade networks in figures 16 to 20 also indicate the centrality of the network or *central betweenness*. *Central betweenness*[51] is a measure of vertex centrality in a graph, i.e. it quantifies the number of times a node acts as a bridge along the shortest path between two other nodes. In practice, it indicates that the municipalities that are being connected with the blue lines are the central municipalities in the network formed in the central and northern arc, in the southern arc it was divided by state because the network is very extensive and for better visualization it was decided to separate by state.

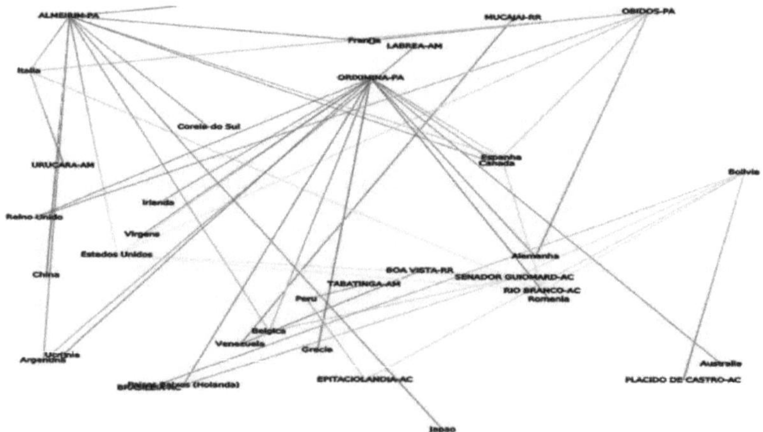

FIGURE 16 - NETWORK FORMED BY FOREIGN TRADE IN THE NORTHERN ARC

---

[50] It should be noted that the database uses the company's tax domicile as a variable to characterize the municipality, i.e. for a legal entity, the tax domicile will be equivalent to the registered office or the place where it carries out its activity. Products do not necessarily have to be manufactured in the same municipality. Simply put, it's where the company issued the invoice to export/import.

[51] See the appendix for more details.

SOURCE: AliceWeb. Own elaboration.

Figure 16 shows the network formed by foreign trade between the border municipalities of the northern arc with other countries. We can see the centrality of the municipality of Oriximiná (PA) and Almeirim (PA) as the main export nodes in the northern arc. In 1999, the export network in the northern arc consisted of 36 nodes and 47 *edges* (lines - which in this case connect the cities to the countries); in 2013, there were 145 nodes and 89 lines. This difference indicates the greater insertion of the central arc into the international trade circuit.

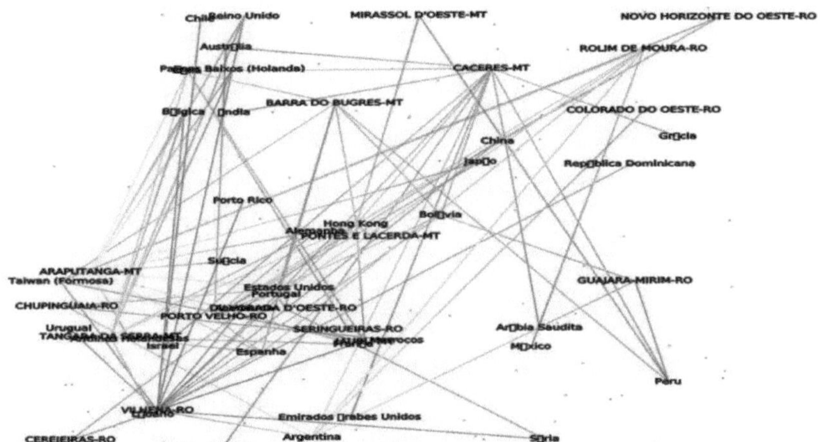

FIGURE 17 - NETWORK FORMED BY FOREIGN TRADE IN THE CENTRAL ARC
SOURCE: AliceWeb. Own elaboration.

Figure 17 shows the network formed by foreign trade between the border municipalities of the central arc and other countries. We can see the centrality of the municipality of Vilhena (RO) and Cáceres (MT) as the main export nodes in the central arc. In 1999, the export network in the northern arc consisted of 145 nodes and 94 *edges* (lines - which in this case connect the cities to the countries); in 2013, there were 521 nodes and 399 lines. This difference indicates the greater insertion of the central arc into the international trade circuit.

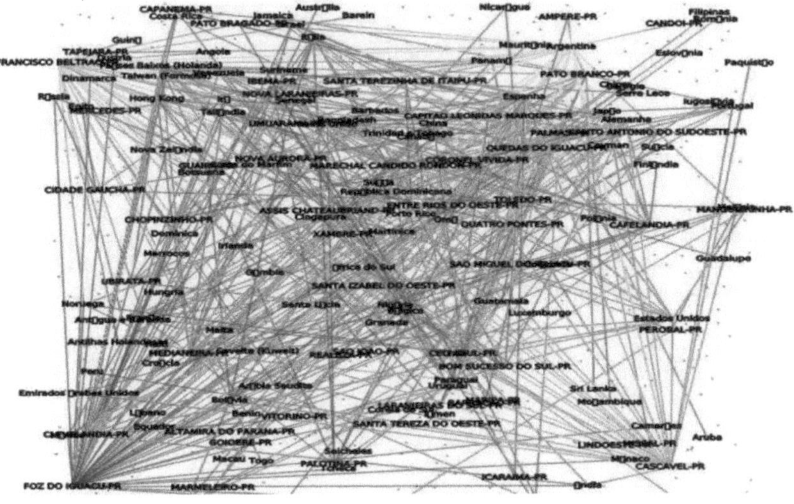

FIGURE 18 - NETWORK FORMED BY FOREIGN TRADE IN PARANÁ
SOURCE: AliceWeb. Own elaboration.

Figure 18 shows the network formed by foreign trade between the municipalities bordering the state of Paraná with other countries.

We can see the centrality of the municipality of Foz do Iguapu (PR) as the main gateway for exporting products. In 1999, the export network in the southern arc consisted of 281 nodes and 1096 *edges* (lines - which in this case connect the cities to the countries); in 2013, there were 2298 nodes and 1962 lines.

In the state of Rio Grande do Sul, the municipality that stands out for its centrality is Erechim (figure 19). What happens on the border of Rio Grande do Sul is an interesting case, as most of the cities in Rio Grande do Sul have a link with Argentina and Uruguay, and what also happens is that these countries in the construction of the network overlap and gain centrality in relation to the other countries.

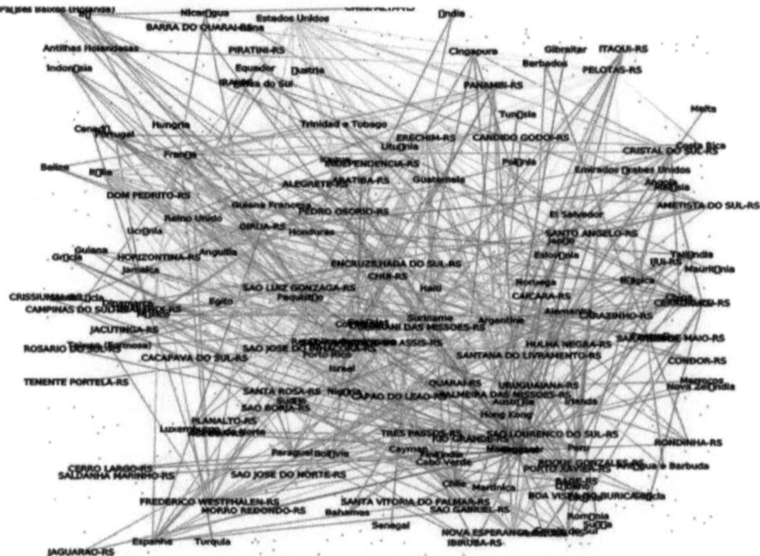

FIGURE 19- NETWORK FORMED BY FOREIGN TRADE IN RIO GREAT SOUTH
SOURCE: AliceWeb. Own elaboration.

Figure 20 shows the network formed by the foreign trade of Santa Catarina's border municipalities. The municipality that stands out for its centrality is Xaxim (SC). Argentina overlaps and gains centrality in relation to the other countries. This means that in the case of Santa Catarina, the link between the territory of Santa Catarina and Argentina is strong, to such an extent that it is possible to measure the centrality of this link. Thus, it can be said that geographical proximity and agreements to form economic blocs are responsible for the centrality of the specific municipalities on the border between Rio Grande do Sul and Santa Catarina and Argentina and Uruguay.

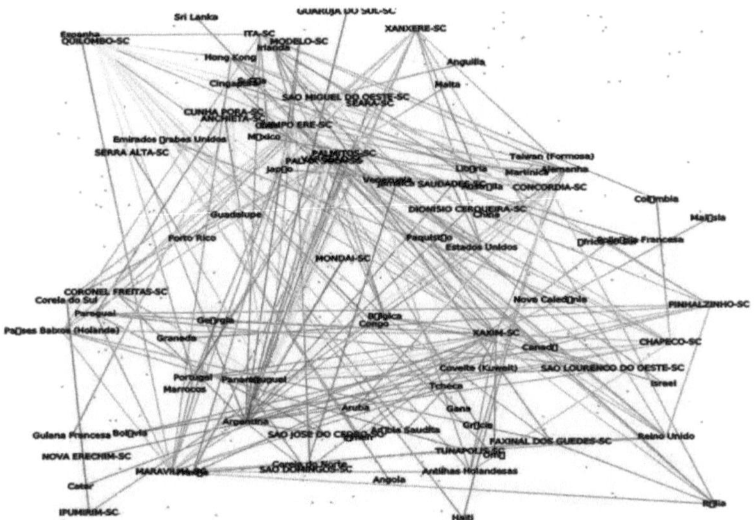

FIGURE 20- NETWORK FORMED BY FOREIGN TRADE IN SANTA CATARINA
SOURCE: AliceWeb. Own elaboration

The municipalities that stand out have flows to several countries, not just South America. The foreign trade flows generated in the border municipalities are multidirectional, i.e. they go to various countries around the world. In 1999, there were a total of 161 countries in Europe, America, Africa and the Middle East. In 2013, the number of countries was 170, also in Europe, America, Africa and the Middle East.

Figures 16 and 17 show the spatialization of exporting cities by border arc, according to the number of countries with which there have been trade relations.

On the other hand, the municipalities that stand out have flows to several countries, not just South America. The foreign trade flows generated in border municipalities are multidirectional, i.e. they go to various countries around the world. In 1999, there were a total of 161 countries in Europe, America, Africa and the Middle East. In 2013, the number of countries was 170, also in Europe, America, Africa and the Middle East.

Table 8.6 shows the relationship between border towns, by border arc, and other countries. It can be seen that the Central Arc is the one that is showing upward growth, and this data, according to Rozenblat (2004), may be an indicator of the multiple social connections that form the cohesion of territories. In these connections, social networks contribute to guiding and channeling intra- and inter-company and inter-country spatial exchanges and transactions. Rozenblat (2004) also states that the spatial dimension of the integration of companies and cities is linked to other strategies that respond to the industrial and governance logics of companies.

Figures 21 and 22 show the spatialization of exporting cities by border arc, according to the number of countries with which there were trade relations.

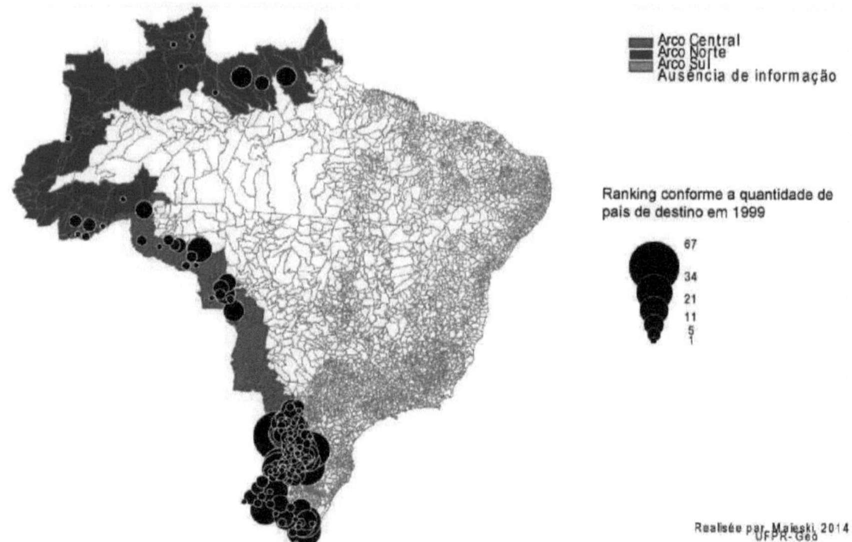

FIGURE 21- RANKING OF MUNICIPALITIES ACCORDING TO THE NUMBER OF
COUNTRIES OF DESTINATION IN 1999
SOURCE: AliceWeb. Own elaboration

According to the *ranking* by order of importance, the most "internationalized" municipalities (which had the most trade relations in relation to the number of countries) in 1999 were: Foz do Iguaçu, Francisco Beltrao, Erechim, Horizontina, Palmas, Rio Grande, Cascavel, Pelotas, Santa Rosa, Toledo, Xaxim and Sant'Ana do Livramento. Due to the commercial dynamism of the Southern Arc, it was expected that the top positions would be occupied by municipalities in this arc, which is what happened. The municipalities of the Southern Arc have a greater international presence than those of the Central and Northern Arcs. The first city in the Central Arc, which appears in 16th[a] position, is Vilhena, in Mato Grosso. In the Northern Arc, the first city in 28th place is Oriximiná, in Pará.

Comparatively, between 1999 and 2008 there was a greater dynamism in foreign trade in the Central Arc, probably driven by the *agribusiness* industry, which demonstrates spatial changes in the border arc and includes it in the international economic circuit. Trade is expanding to 60 new countries, reflecting the search for new markets, mainly in the Middle East and Asia. The Central Arc is gaining greater positions in terms of the number of countries with which it has trade relations. In order of importance, the border towns are: Erechim, Foz do Iguaçu, Rio Grande, Cascavel, Hulha Negra, Xaxim, Bagé, Horizontina, Capao do Leao, Palotina, Palmas, Porto Velho and Vilhena.

Porto Velho, located in the Arco Norte, is ranked 11th and Vilhena 12th. Between 1999 and 2008, the Northern Arc lost importance in order of the number of exporting countries. Rio Branco is the first, appearing in 48th place.

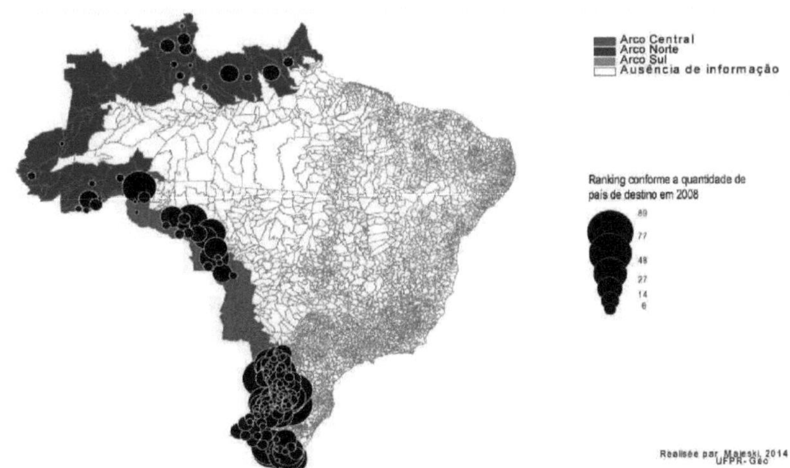

FIGURE 22- RANKING OF MUNICIPALITIES ACCORDING TO THE NUMBER OF
COUNTRIES OF DESTINATION IN 2008
SOURCE: AliceWeb. Own elaboration

In 2013, the cities in order of importance were: Rio Grande, Cascavel, Foz do Iguapu, Erechim, Palmas, Vilhena, Sapezal, Palotina, Seara, Santa Rosa and Hulha Negra. The first cities in the Central Arc are Vilhena and Sapezal, which exported to 54 and 51 countries respectively. In the Northern Arc, the first city is Almeirim, in 42ª position.

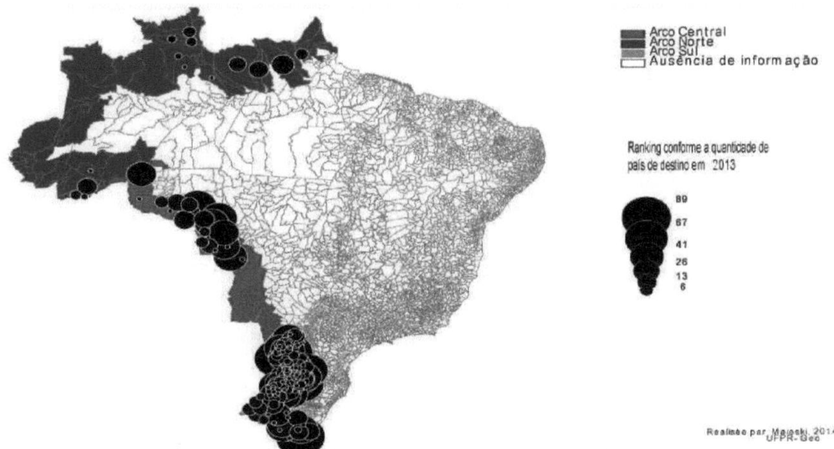

FIGURE 23- RANKING OF MUNICIPALITIES ACCORDING TO THE NUMBER OF
COUNTRIES OF DESTINATION IN 2013

There was a clear intensification of multidirectional flows in the three border arcs between 1999 and 2013. The search for new markets, tax incentives and new agreements between economic blocs have all helped to insert these cities into the international economic circuit. Table 8.6 shows the number of *links in* each border arc in 1999, 2008 and 2013. *Links* are connections between cities and with countries. This forms part of the international trade network, which can only be identified from city to country, according to the data available in the MDIC database.

According to the hierarchy of exporting companies on the border, Foz do Iguaçu, Cascavel, Porto

Velho, Rio Grande, Pelotas, Chapecó, Erechim, Boa Vista, Ponta Pora, Guajará-Mirim, Barracao, Porto Xavier and Umuarama are the main cities with the largest number of exporting companies. Figure 24 shows the spatial distribution of exporting companies and figure 25 of importing companies according to geographical location. In the reverse flow (imports), the largest importing cities are Rio Branco, Erechim, Boa Vista, Barracao, Porto Xavier, Toledo, Uruguaiana, Umuarama, Ponta Pora, Panambi, Dourados and Francisco Beltrao.

There is a concentration of exporting companies in the Southern Arc, as it is the most dynamic in terms of population, companies and number of municipalities. Figures 26 and 27 detail the number of companies in the Southern Export and Import Arc.

FIGURE 24- AVERAGE NUMBER OF COMPANIES EXPORTERS
SOURCE: MDIC . Own elaboration

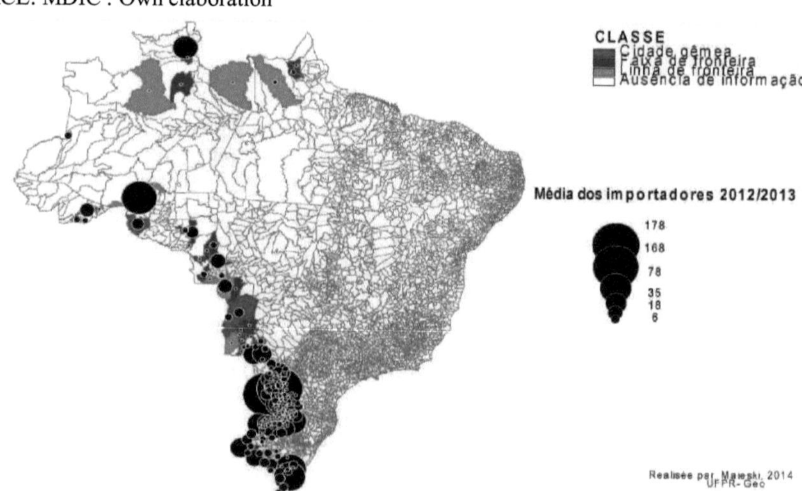

FIGURE 25-AVERAGE NUMBER OF IMPORTING COMPANIES
SOURCE: MDIC . Own elaboration

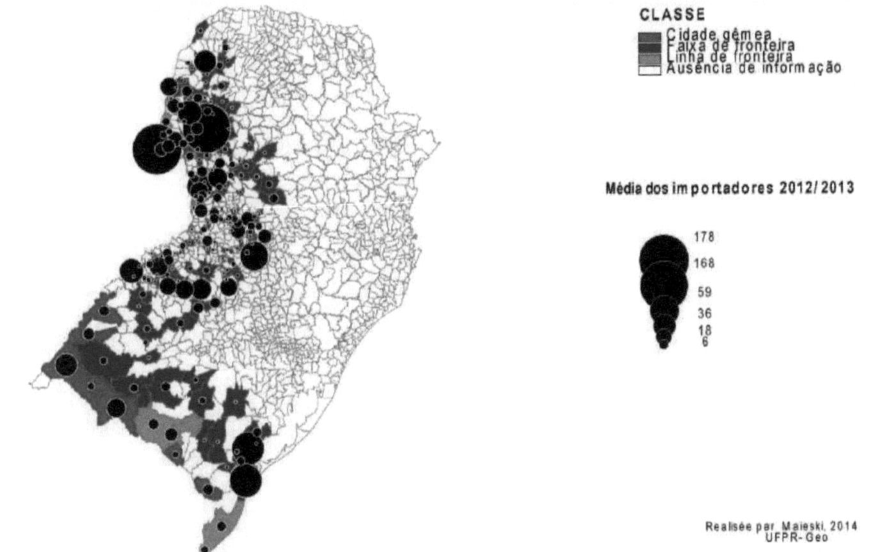

FIGURE 26-AVERAGE NUMBER OF IMPORTING COMPANIES IN THE SOUTHERN ARC
SOURCE: MDIC . Own elaboration

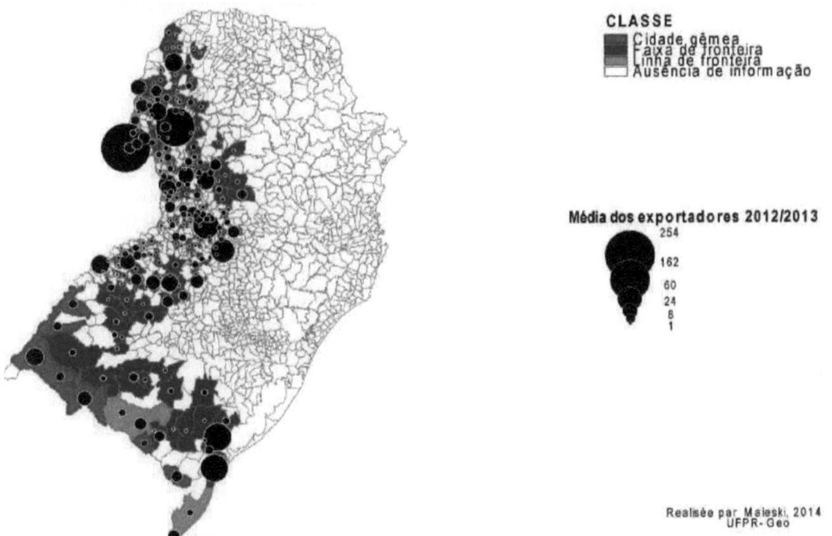

FIGURE 27-AVERAGE NUMBER OF EXPORTING COMPANIES IN THE SOUTHERN ARC
SOURCE: MDIC

Companies located in a city are part of a set of urban dimensions and infrastructure that are fundamental to the flow of international trade. The twin cities of Rio Grande do Sul stand out for their logistical importance, especially those bordering Uruguay. They are home to many exporting companies and serve as a logistics platform for international trade.

Table 14 shows the number of exporting/importing companies according to geographical location on the border strip. There is a concentration of companies in the municipalities located on the border strip (which do not touch the international boundary).

TABLE 14 - GEOGRAPHICAL LOCATION OF EXPORTING AND IMPORTING COMPANIES

CLASS	Exporters	Importers
Gem City	647	496
Border strip	1492	1232
Border line	255	222
Grand total	2394	1950

Exporting companies are concentrated in the Southern Arc, with 77% located there, 4.5% in the Northern Arc and 18.5% in the Central Arc. Importing companies are located 79% in the South, 4.7% in the North and 16% in the Central Arc.

Despite the privileged geographical location of the border towns, there are difficulties in integrating them into an international trade context, which can be explained by various factors, such as: bureaucratic and legal obstacles, lack of agility in the services provided at the head of borders, disparity in prices between markets, restrictions on the transit of retail goods, unilateral government decisions and differences in customs treatment when importing and exporting products (MENDONQA, 2009). In addition to these factors, there is the issue of safe international logistics, which is becoming increasingly important worldwide, as Morini and Leoce (2011) point out, presenting a World Bank study which places Brazil in an intermediate position[52] in terms of logistical efficiency.

---

[52] The World Bank's Logistics Efficiency Index (LPI) ranks Brazil 41[a] out of 155 countries surveyed. If only the item "customs administration" is highlighted, Brazil moves up to 82 [a](http://www.worldbank.org).

CHAPTER 7

This chapter aims to conclude the discussion on the transition from the defense paradigm to defense and development. The chapter is dedicated exclusively to the issue of border defense at the present time. It discusses the role of the Integrated Border Monitoring System (SISFRON), an audacious project underway by the Brazilian Army, which recently began operations (second half of 2014).

SISFRON is an integrated sensing, decision support and operational deployment system whose purpose is to strengthen the state's presence and capacity for action on the border strip. The sensors will be distributed along the 16,886 kilometers of the FF for 10 years, linking the military commands of the Amazon, West and South.

In addition to monitoring, the main objective is to ensure a reliable and timely flow of information for decision-making, to act in defense actions or against environmental and cross-border crimes. This information will be processed in decision-making centers that will have to rely on highly integrated and reliable command and control systems. SISFRON aims to improve the Army's structure and encourage the defense industry, integrating the Armed Forces' sister systems. In addition, it is an institutional instrument because it integrates various levels of land force deployment (battalions, brigades, divisions, military commands).

7.1 THE TRANSITION FROM THE DEFENSE PARADIGM TO THE SECURITY AND DEVELOPMENT PARADIGM

Following the promulgation of the 1988 Federal Constitution, the paradigm applied to the border strip from a military point of view changed from an area of security to one of defense and development (as discussed in Chapter 4). Programs such as Calha Norte were set up to develop and defend the Amazon region, among other functions. One of the most recent and most audacious projects is SISFRON, which aims to monitor the entire border strip, i.e. 27% of the territory, in order to help combat environmental and border crimes. In this vision, the border strip becomes an important object of integration with neighboring countries in order to combat cross-border crimes that occur at certain points along the border.

Political interest in building economic blocs and South American integration also influenced the transformation of the Armed Forces. In 1999, the Ministry of Defense was created, marking the third change in the military paradigm. From a military point of view, the border area became one of defense and development. In 2005, the National Defense Policy document was published and in 2008, the National Defense Strategy (END). These two documents are the most important on the subject because they provide for military cooperation with South American countries in the field of defense.

In 2004, a complementary law was also published on the Army's role on the border strip, authorizing it to act through repressive and preventive actions against cross-border and environmental crimes, together with other agencies. The Army began to direct its practices to the border strip, mainly in relation to preventive and repressive actions against cross-border and environmental crimes. It can act alone or in coordination with other executive bodies, mainly in patrol activities, searches of people and land vehicles, boats and aircraft, and arrests in flagrante delicto (Law No. 97 of June 9, 1999 and Complementary Law No. 117 of September 1, 2004) (FURTADO, 2013).

In particular, the END is organized around three main points, called structuring axes: reorganization of the Armed Forces, reorganization of the national industry and defence material, and composition of the Armed Forces' personnel.

SISFRON is part of this context. It is an integrated sensing, decision-support and operational deployment system whose purpose is to strengthen the state's presence and capacity for action along the border. In addition to monitoring, the main objective is to ensure a reliable and timely flow of information for decision-making, to act in defense actions or against cross-border crimes. In short, the system is based on the production and dissemination of information gathered by sensors strategically installed along the Brazilian land border. This information will be processed in decision-making centers that will have to rely on highly integrated and reliable command and control systems.

SISFRON aims to improve the structure of the army and stimulate the defense industry, integrating the armed forces' joint system. The sensors will be distributed along the 16,886 kilometers of the FF linked to the military commands of the Amazon, west and south. In addition, it is an institutional tool because it integrates various levels of land force deployment (battalions, brigades, divisions, military commands).

An important aspect is SISFRON's integration with other resources that already exist or are under development, such as the Amazon Protection System (SIPAM), the Blue Amazon Management System (SISGAAZ) and the Airspace Control System. SISFRON works together and feeds data and information into the databases of the Federal Police, Federal Highway Police, Civil Police, Military Police and municipal, state and federal inspection bodies. The expectation is to be able to respond in real time to detected problems.

7.2 HOW SISFRON WORKS

SISFRON is structured in three main parts: sensing, actuators and action. The subsystems are: sensing, decision support, action, communication, information security, simulation and human resources and logistics capacity (Figure 10.1).

1. Sensing subsystem: this is the collection of images and information on the flows that cross the border. The subsystem is made up of a set of radars, ground surveillance, optical sensors and electromagnetic signal sensors. The sensors will be fixed or mobile and installed on vehicles, specialized vessels, special platforms, satellites and fixed-wing aircraft. The project envisages the production of 20 and 200 km VANS as a monitoring tool. The sensors will be responsible for the constant monitoring of areas of interest in the national territory, particularly along the border.

2. Decision support: processing the data collected by sensors. The aim is to process the information they generate. The decision-maker will be supported by a fixed information network, a support network and an operations network. At this stage, the information from the human, image, electronic and signal sensor systems is analyzed. The information undergoes intelligence analysis and technical data processing. This phase is important for determining what will be done at the operational level to combat crime at the border. Decision-makers work on analyzing and processing the data collected, comparing the action options that emerge and, in due course, making appropriate and relevant decisions.

3. Information and communication technology subsystem: carries out data traffic between SISFRON components. It is the fixed and mobile communications network.

4. Information and communication security subsystem: ensures the reliability, authenticity and availability of system information.

5. Simulation and training subsystem: aimed at training and specializing the personnel involved in the management and operation, logistics and development of the system.

6. Logistics subsystem: comprises the infrastructure needed to carry out maintenance, supply and transportation activities.

7. Actors subsystem: comprises the actions of military forces and agencies to combat border crimes. The role of the actors is to provide a prompt response to the problem detected, relying on strategic mobility to achieve this goal.

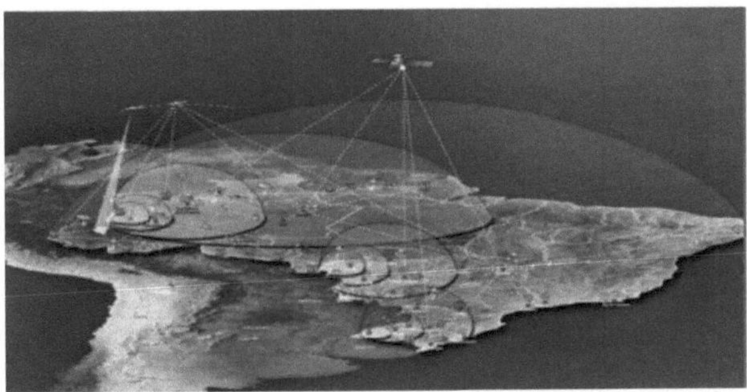

Figure 10.1 - How SISFRON works
source: http://www.defesanet.com.br/terrestre/noticia/1651/SISFRON--- Border-Protection-and-the-National-Defense-Industry--Informe- COMDEFESA)%20[between

SISFRON envisages networking with other government bodies, including linking up existing systems such as SIPAM, which monitors the Amazon. The project has an estimated cost of around 12 billion and is expected to be fully implemented by 2026. The pilot project is being installed in the state of Mato Grosso do Sul on the border with Paraguay, at the 4ª Brigade of Mechanized Cavalry, in Dourados. Today, the Army's presence on the border is still limited due to the length of the border (Figure 7.2).

Figure 7.2 - Brazilian Army presence on the border strip

The choice of location for the pilot project is justified because starting such a project in the North would be very costly and difficult. Difficult because there are few land links. Costly because it is necessary to develop more advanced technology to monitor crimes, taking into account the vegetation cover.

The southern region doesn't have the most serious problems. It has a good population density and good logistics and is not the Army's main focus at the moment.

The western region was chosen because it has a high rate of cross-border and environmental crime, due to the permeability of the dry border. In addition, the Western Military Command has military units spread over almost the entire border line. The state of Mato Grosso's road network also contributed to this choice.

The system will be expanded gradually. Initially in the northern states of Mato Grosso and Rondônia, then in the southern states of Paraná and Santa Catarina. In a third stage, it will be extended to the states of Rio Grande do Sul, Acre, Amazonas, Roraima, Pará and Amapá.

7.3 FACTORS FOR THE CREATION OF SISFRON

In a nutshell, it can be said that the high homicide rate, the incidence of drug trafficking, human trafficking, geopolitical issues and, above all, the political will to develop a national defense industry were the main factors that motivated the creation of SISFRON.

The geopolitical issue surrounding the SISFRON project is the protection of the Amazon, for example, the long stretch of land border with the Amazon Rainforest, the largest tropical rainforest in the world. The pan-Amazon is present in 9 South American countries and occupies around 40% of Brazil's territory. The forest is an important water reserve with a great biodiversity of fauna and flora.

The Brazilian Army considers the Amazon region to be a vulnerable area due to the forest's environmental conditions and low population density.

7.4 HOMICIDE RATE

The Ministry of Health database was used to calculate the homicide rate in border municipalities. The rate was calculated using the average for the years 2009 to 2011 for municipalities with more than 20,000 inhabitants. The result is a homicide rate per 10,000 inhabitants. Figure 7.2 shows that the most critical region is the central border, which was chosen to host the pilot project. Although the study focuses on border municipalities, this does not mean that homicides only occur on the borders.

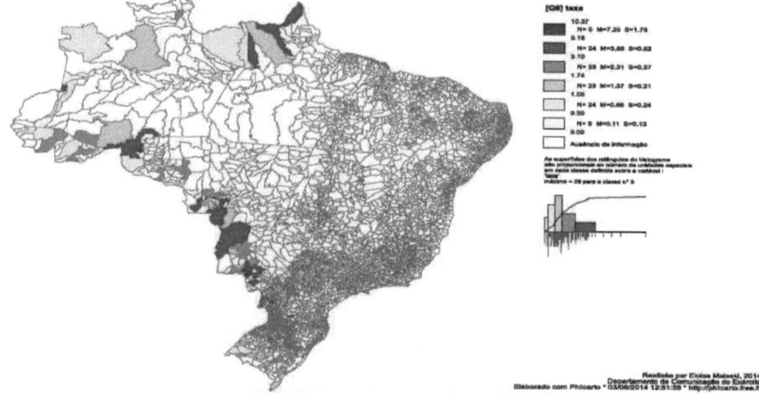

Figure 7.2 - Homicide rate per 10,000 inhabitants between 2009 and 2011

Graph 7.1 shows the relationship between the homicide rate and the urban hierarchy. The study shows that the homicide rate does not respond to a law of demographic scale. The cities of Foz do Iguapu, Porto Velho and Cascavel have an average homicide rate higher than the ratio between cities.

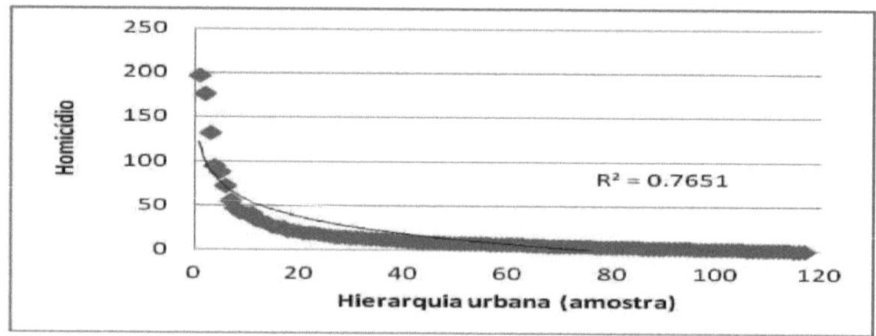

Graph 7.1 - Homicide rate and urban hierarchy in border municipalities

The Army estimates that the cost of violence in Brazil in 2010 was around 5.09% of GDP (183.5 billion). The high homicide rate may have something to do with drug trafficking.

Brazil shares a border with cocaine producers and, according to a UN report, the length of the border and its proximity to the largest cocaine producers, together with the large population and the length of the sea coast, make Brazil a transitory country on the trafficking route. In 2011, the cocaine seized in Brazil originated in Bolivia (54%), followed by Peru (38%) and Colombia (7.5%). Since Bolivia has no access to the sea, it identifies Brazil as a transit destination. When cocaine enters Europe, South America appears several times as a transit country.

Another problem related to illicit flows concerns public security. According to the project to create SISFRON, it is estimated that "approximately
400,000 vehicles and 15,000 loads and a large part of these vehicles and loads are taken out of the country across borders. Every year, there are 125,000 seizures of narcotics and 80,000 seizures of firearms, and most of the narcotics and firearms seized enter Brazil through the borders. Finally, it should be noted that approximately 33,000 people go missing in Brazil every year and a large proportion of them are taken out of Brazil through these borders."

7.5 EXPECTATION OF RESULTS

SISFRON aims to monitor and combat illicit border flows. The bill also seeks to encourage and develop the national defense industry. The law gives preference to Brazilian companies for the development of equipment and software to be used in SISFRON. The expectation is to generate direct and indirect jobs around the system. Table 7.1 shows the expected number of jobs over the years when the system is deployed.

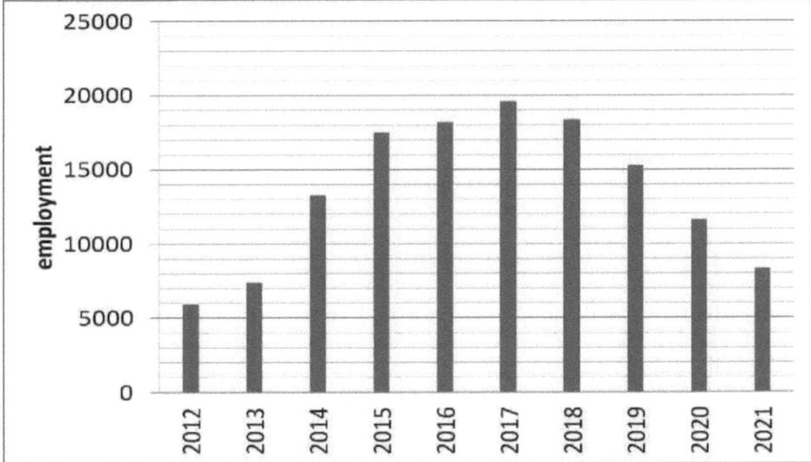

Table 7.1: Job creation expectations

It is estimated that 2/3 of the jobs could come from the technology sectors, carrying out component and subsystem development activities, software development and system integration by national companies.

A second improvement would be the net contribution to the trade balance resulting from exports of products developed or manufactured in the country for SISFRON. The contributions of radars, UAVs and command and control system projects were considered.

As addressable markets, only Latin America (excluding Brazil) and Africa were considered, areas of greater influence for Brazil and the potential national exporters involved, during the SISFRON operating timeframe. The deployment period was taken into account as the system's exposure period and, based on the results achieved, the leverage of opportunities for future sales of this equipment to other countries.

REFERENCES

ALICEWEB - **Foreign Trade Information Analysis System.** Municipal exports.
MDIC, Brasilia. Available at: http://aliceweb.desenvolvimento.gov.br/
ALICEWEB - **Foreign Trade Information Analysis System.** Municipal imports.
MDIC, Brasilia. Available at: http://aliceweb.desenvolvimento.gov.br/
ALICEWEB - **Foreign Trade Information Analysis System.** Brazilian exports.
MDIC, Brasilia. Available at: http://aliceweb.desenvolvimento.gov.br/
ALICEWEB - **Foreign Trade Information Analysis System.** Brazilian imports.
MDIC, Brasilia. Available at: http://aliceweb.desenvolvimento.gov.br/
ALMEIDA, PAULO ROBERTO DE. Brazil and regional blocs: sovereignty and
interdependence. **Sao Paulo Perspec.**, Sao Paulo , v. 16, n. 1, Jan. 2002 .
ALMEIDA, P.R.**Formapáo da diplomacia económica no Brasil : as relapóes
económicas internacionais no Império.** Publisher, SENAC: Brasília, FUNAG,2001.
ALMEIDA, P. R. Conjunctural and structural problems of integration in South
America: the trajectory of Mercosur from its origins to 2006. **Meridiano 47** - Boletim
de Análise de Conjuntura em Relapoes Internacionais. N. 68, p 4-10. 2006.
ANTUNES, E. M. **Revisáo** dos **critérios de definipáo de cidades-gemeas nas
fronteiras brasileiras**.Disponível em: http://www.jornal.ceiri.com.br/revisao-dos-
criterios-de-definicao-de- cidades-gemeas-nas-fronteiras-
brasileiras/../../Documents/CEIRl/revisao- dos-criterios-definicao-de-cidades-gemeas-
nas-fronteiras-brasileiras.pdf Accessed 13/08/2014.
AVEIRO, T.M.M. **Brazil-Uruguay Relations: The New Agenda for Cooperation
and Development
Fronteiripo.**(Dissertation).UNB,Brasilia, 2006.
CENTRAL BANK OF BRAZIL - BCB. **Foreign exchange.** Available at:
http://www.bcb.gov.br/7cambio.acesso on 24/05/2012.
BANDEIRA, L.A M. O Baräo de Rothschild e a questão do Acre. **Rev. bras. polít. int.,**
Brasília, v. 43, n. 2, Dec. 2000.
BARRAL, W.; BOHER, C. P. **Latin American integration in focus: 50 years of
ALALC/ALADI.** In: Latin American integration: 50 years of ALALC/ALADI.
Brasília: FUNAG, 2010.
BARROSO, L.A. A faixa de fronteira: procedimento ratificatório de titulapöes. **Revista
Fac. Direito**, v.19/20, n1, p.19-28, jan/dez,1995/1996.
BORBA, v. Fronteiras e faixa de fronteira: expansionismo, limites e defesa. **Historia,**
Rio Grande, v. 4, n. 2: 59-78, 2013.
Brazil, Ministério da Integrapäo nacional (MI), **Programa de Desenvolvimento da
Faixa de Fronteira: Proposta de Reestruturagao do Programa de Desenvolvimento
da Faixa de Fronteira.** Brasília, MI/SPR, 415 p., 2005.
BREITBACH, A. Between industrial specialization and diversification: towards
sustainable regional development. **Perspectiva Económica On-line,** 1(2): 1-30,
Jul./Dec. ISSN 1808-575X, 2005.
CARNEIRO FILHO, C.P. **Interaçoes espaciáis e cidades-gêmeas na fronteira Brasil-
Argentina: Sao Borja/Santo Tomé - Itaqui/Alvear.** Available at:
http://www.retis.igeo.ufri.br/. Accessed on 02/05/2012.
CARVALHO, G.B. MOURÄO, M. **A vertiginosa espiral da racionalidade.** In: Obras
do Barao do Rio Branco II : questões de limites Guiana Inglesa. - Brasilia: Alexandre de
Gusmao Foundation, 2012. 214 p.; ISBN 978-85-7631-353-3
CATELA,E.Y.S. (2009) **Essays on International Trade, Technology and Growth.**
Thesis (Doctorate). Postgraduate Program in Economic Development. UFPR: Curitiba,
2009.
CEDEPLAR.Critical publication of the 1872 general census of the empire of brazil.
Center for Research in Economic and Demographic History - NPHED. 2012.

Available at
http://www.nphed.cedeplar.ufmg.br/wp- ontent/uploads/2013/02/Preliminary report
1872 site nphed.pdf .
CNM - National Confederation of Municipalities. **Final report of the First Meeting of Border Municipalities,** 2008.
COELHO, P.M. P. **Fronteiras na Amazônia: um espaço integrado**. Alexandre de Gusmao Foundation FUNAG: Brasília, 1992.
CORREA, L. F.S.. The Baron of Rio Branco, Head of Mission: Liverpool, Washington, Bern and Berlin". In: *Barão do Rio Branco - one hundred years of memory*. Brasília: FUNAG, p. 263-79, 2012.
COSTA, W.M. Brazil and South America: geopolitical scenarios and the challenges of integration. **Confins**. Number 7, 2009.
COSTA, W M. Políticas territoriais brasileiras no contexto da integraçao sul-americana. **Revista Território**, Rio de Janeiro, ano IV, n 7, p. 25-41, iul/dez, 1999.
COUTO, L. F. **O Horizonte regional do Brasil: Integragáo e construgáo da América do Sul.** Curitiba: Juruá, 2009.
DATHEIN,R. Economic integration and development policies: experiences and perspectives. **Revista Análise Económica,** Porto Velho, year 25, p. 49-69, September 2007.
DORATIOTO, F. Rio Branco and the Palmas Question. In: **Obras do Baráo do Rio Branco I : questões de limites República Argentina.** - Brasília: Fundapao Alexandre de Gusmao, 2012.
Doratioto, F. **O Brasil no Rio Prata (1822-1994)** Ed. - Brasília: FUNAG, 2014.
EGLER, A.G. C. **Economic Integration and Logistics Networks in the Southern Cone.** University of Brasília, pg. 312-326. Available at: ttp://egler.net/index.php/ensitec?view=publication&task=show&id=10, 2001.
FARIA, L.A, COUTINHO, C.R. **Trade relations and integration in South America.** Available at: http://www.fee.tche. br/sitefee/download/tds/084. pdf, 2009.
FERREIRA, A. C. Interactions on the Brazil-Uruguay border: a case study of the cities of Jaguarao-RS (Brazil) and Río Branco (Uruguay).
Boletim do TEMPO Electronic Magazine, Year 4, N°37, Rio, 2009.
Foucher, M. **L'obsession des frontières.** Paris/Librairie Académique Perrin, 248p., 2007.
FOUCHER, M. **Fronts et frontières : un tour du monde géopolitique.** Libraire Arthème Fayard, 1991.
FOUCHER, M. **L'obssession des frontières.** Perrin Publishing. Tempu Collections. Paris, 1991.
FOUCHER, M. **La réaffirmation mondiale de frontières.** Editora Armando Colin, pg. 23-30, 2011.
Furtado, R. **Descobrindo a faixa de fronteira: a trajectória das elites organizacionais do Executivo federal: as estratégias, as negociaçoes e o embateado na Constituinte.** Curitiba, CRV, 350 p., 2013.
GÉOCONFLUENCES. **General notions of programs.** Available at: http://geoconfluences.ens-lyon.fr/. Accessed 03/03/2013.
GREMAUD, A.P.. VASCONCELLOS, M. A S, TONETO JUNIOR, R. **Contemporary Brazilian economy.** 7. ed. Sao Paulo (SP): Atlas, 2007
HAESBAERT, R.. BARBARA, M.JS. Identity and Migration in Cross-Border Areas. **GEOgraphia**, Vol. 3, No 5 (2001). Available at: http://www.uff.br/geographia/ojs/index.php/geographia/article/view/53
HARVEY, D. **The enigma of capital and the crises of capitalism.** Sao Paulo, SP: Boitempo, 2011.

IBGE - Brazilian Institute of Geography and Statistics. **Demographic census 1872-2010.** Available at: www.ibge.gov.br Accessed on January 11, 2014.

IBGE - Brazilian Institute of Geography and Statistics. *IBGE Cities.* Available at : www.ibge.gov.br Accessed on January 11, 2014.

IIRSA. **The Initiative for the Integration of South American Regional Infrastructure.** Available at: http://www.iirsa.org/. Accessed 02/02/2011

LAFER, C. Discurso: **Desenvolvimento e Integragáão na América Latina e no Caribe: a Contribuido das Ciencias Sociais.** MRE: Brasília, 2002.

Lessa, A.C. O Barao do Rio Branco e a inserpão internacional do Brasil. *Rev. bras. polít. int.* [online]. vol.55, n.1, pp. 5-8, 2012.

LIMA, FR F. **Desenvolvimento regional na fronteira Foz do Iguapu/BR - Ciudad Del Este/PY.** Doctoral thesis, UFPR: Curitiba, 2011.

LIMA, S L M. **O acompanhamento tributário - um novo paradigma em fiscalizapao para a Receita Federal do Brasil.** ESAF - School of Treasury Administration. Award-winning monograph. Brasilia, 2007.

LOBO. **Economic geography.** 12 ed. Säo Paulo: Atlas, 1997.

Marques, A M. **Border migratory movements: Bolivians and Paraguayans in Mato Grosso do Sul.** Available at: http://lanic.utexas.edu/project/etext/llilas/ilassa/2007/marques.pdf

MATTOS, C.M. **Geopolítica e teoria de fronteiras do Brasil.** Rio de Janeiro: Army Library, 1990.

MATTOS, C.M. **Uma geopolítica amazónica.** Rio de Janeiro: Army Library, 1980.

MATTOS, C.M. **Geopolitics and destiny.** Rio de Janeiro: Army Library, 2000

MATHIAS, S Kl; GUZZI, A C; GIANNINI, R. A. Aspects of regional integration in defense of the Southern Cone. **Revista Brasileira Política Int.** 51 (1): 70-86 (2008).

MDIC. **200 years of foreign trade.** Available at: www.mdic.gov.br . Accessed 2010.

MDIC. **Terms of Reference: Internationalization of Brazilian companies.** MDIC, Brasília, 2009.

MENDONÇA, C. Impacts of integration processes in border areas: the growth of trade flows and development in twin cities of Mercosur. **CEPPAC SERIES - Center for Research and Postgraduate Studies on the Americas:** Brasília, 2009.

MENEZES, A.M, PENNA FILHO, P. **Integraçao regional: blocos económicos nas relações internacionais.** Rio de Janeiro, Elsevier, 2006.

MERCOSUR. **General aspects of Mercosur.** Available at: http://www.mercosul.gov.br/. Accessed 20/02/2013.

MINISTRY OF NATIONAL INTEGRATION. **Border Strip** Development Promotion Program - PDFF. SPR . Secretariat for Regional Programs: Brasília, 2009.

MINISTRY OF NATIONAL INTEGRATION. **Proposal for Restructuring the Border Strip Development Program.** Brasília: Ministry of National Integration, 2005.

MOREIRA, A. **Teoria das Relaçôes Internacionais.** Porto: Almedina, 3 ed. pg. 487-512, 2003.

MORINI, C.; and LEOCI, G. **Secure international logistics: Authorized Economic Operator and border management in the 21st Century.** Sao Paulo: Atlas, 2011.

MRE - Ministry of International Relations. *Works of Barao do Rio Branco I : boundary issues Argentine Republic.* - Brasília: Alexandre de Gusmao Foundation, 295p., 2012.

OLIVEIRA, R,G MAGALHÄES, M.G Question of the Pirara: Roraima. 2002. **Texts & Debates,** Available at http://revista.ufrr.br/index.php/textosedebates/article/viewFile/878/723.

OLIVEIRA, J. L. C. **Manaus Free Trade Zone: a study on the renunciation of federal entities and the socio-economic benefits generated.** Master's dissertation.

Porto Alegre: UFRGS, 2011.

PERROUX, F. **Essay on the philosophy of new development**. Lisbon: Calouste Gulbenkian Foundation, 1981.

PORTO, J.; COSTA, M. **A Área de Livre Comércio de Macapá e Santana: Questöes Geoeconômicas**. Macapá: Editora O Dia, 1999.

PATRIOTA, A. A. Preface. In: **Obras do Barao do Rio Branco I : questões de limites República Argentina.** - Brasilia: Alexandre de Gusmao Foundation, 2012.

UN. **World Drug Report 2013**. UNITED NATION, New York, 2013.available at: http://www.unodc.org/lpo-brazil/en/drugs/world-report-on-drugs.html

PRAZERES, T.L. **South American integration: an idea still out of place?** In: Brazil and South America: challenges in the 21st century / Brasilia: Fundapao Alexandre de Gusmao: Instituto de Pesquisa de Relapóes Internacionais, 2006. 150p. ISBN 85-7631-060-0

PORTO, J.L R. **Amapá: Main economic and institutional transformations - 1943-2000.** Macapá: SETEC, 2003.

, et.al. **From Federal Territory to State: conditions for the implementation of spatial adjustments in Amapá.** In: Seminar on Thirty-Five Years of Colonization in Amazonia. Porto Velho (RO), 2007.

PUMAIN, D. SAINT JULIEN, Thérèse. **L'analyse spatiale : localisations dans l'espace** . Armando Colin Publishing House, Paris, 2010.

Pumain D. Pour une théorie évolutive des villes. In: **Espace géographique**. Tome 26 n°2, 1997. pp. 119-134.

FEDERAL REVENUE OF BRAZIL - RFB. **Who are we?** Available at: http://www.receita.fazenda.gov.br/ accessed on 12/12/2012.

REITEL, Bernard. **Frontiers**, ZANDER, Patrick. 2013. Available at : http://www.hypergeo.eu/IMG/ article PDF/article 16.pdf.

REITEL, Bernard. **Frontières**.2004.Available at : http://www.hypergeo.eu/IMG/ article PDF/article 16.pdf.

REZEK, F. **Direito Internacional Público**. Editora Saraiva, Sao Paulo: 2010.

ROCHA, M. S. **Brazil and South America: challenges in the 21st century.** Brasilia: Fundapao Alexandre de Gusmao: Instituto de Pesquisa de Relapóes Internacionais, P. 113-146., 2006.

ROZENBLAT. **Tissus de villes. Urban networks and systems in Europe**. Available at **http://my.unil.ch/serval/document/BIB 597A8335C98F.pdf, 2004.**

SASSEN, S. The repositioning of cities and urban regions in a global economy: expanding policy and governance options. **Revista eure** (Vol. XXXIII, N° 100), pp. 9-34. Santiago de Chile, December 2007.

SCHULZ, C. **L'agglomération Saarbrucken - Moselle -est : ville- frontière ou villes frontalières** in : Villes et Frontières. pg. 50 a 62. Collection Villes Anthropos, 2002.

SEBRAE. **Internationalizing the company**. Sao Paulo, SP: SEBRAE, 2013.

SENHORAS, E M. Dinámica fronteiripa das cidades-gemeas entre brasil e guyana. **REVISTA GEONORTE**, Special Issue 3, V.7, N.1, p.10771094, 2013 (ISSN - 2237-1419).

SENHORAS, E. M. **The New International Regionalism from Theory to Practice: A Case Study of Regional Integration and Panregionalism at the South American Crossroads**. In: VI Mercosur Forum, 2007, Aracajú. Proceedings of the VI Mercosur Forum. Aracajú: UFS, 2007.

SILVA, R.M.; OLIVEIRA, T.C.M. The merit of twin cities in border spaces. Revista **OIDLES** - Vol 2, N° 5, diciembre, 2008.

SILVA, L.P. B. **Productive chains in a border area: Corumbá (MS) and Puerto Suarez (BOL).** In: Anais XVI Encontro Nacional dos Geógrafos: crise, práxis e autonomia: espaços de resistência e de esperanças. Porto Alegre, 2008.

SILVA, L. Paraguay: inconclusive transition and reticent integration. In: **South American Agenda. Changes and challenges at the beginning of the 21st century**. Brasília: FUNAG, 2007.

SIMOES, R; MORINI, C F. The world economic order: considerations on the formation of economic blocs and Mercosur. Revista: **Impulso**, n. 31, 139-154, 2002.

Souza, C.A; Porto, J.L., Pedro, J. M. Santos, M.M. **Território Federal no direito brasileiro: estudo comparativo dos Projeto de Lei 008/1947 e Decreto-Lei 411/1969.** Available at http://www.naea.ufpa.br/siteNaea35/anais/html/geraCapa/FINAL/GT9- 261-1208-20081125194314. pdf accessed on 13/08/2014.

SOUZA CRUZ, G. A.; BISPO, R. S.; SILVA, A. Z. B. "A criação de Zonas de Processamento de Exportaçao e de Áreas de Livre Comércio como instrumentos de redu redução do desequilíbrio intra-regional na Amazônia Ocidental". **Revista Examapaku**, vol. 1, n. 1, 2008.

STEIMAN, R. ; MACHADO, L. **Limits and international borders: a historical and geographic discussion".** Article available at http://www.igeo.ufrj.br/fronteiras . (2002)

SUFRAMA. "ZFM's tax and social gains outweigh tax waivers". **Suframa Hoje**, year X, n. 45, August 2009.

TREVISAN, R. **Newton's Law applied to foreign trade.** Personal interview conducted on 15/12/2012.

UNASUR. **Union of South American Nations**. Available at: http://www.unasursg.org/ Accessed on 01/02/2015.

UTEPI. **Business in Paraguay: elements of the country cost. United Nations Development Organization Cooperation.** Available at: http://www.mic.gov.py/images/costo pais paraguay.pdf. Accessed on 12.Dec.2012.

VAZ, A.C. **Cooperation, integration and the negotiating process: the construction of MERCOSUR.** Barâo do Rio Branco Institute. Brasilia: IBRI, 2002.

VAZ, A C. Mercosur at ten years: growth crisis or loss of identity? **Rev. bras. polít. int.**, Brasília, v. 44, n. 1, June 2001.

VIANA, A R, SILVA, P. Barros, ANDRE, B C. **South American integration: opportunities and challenges for the continent's greater participation in global governance.** In: Global governance and South American integration. Brasilia: Ipea, 2011.

VIZENTINI, P G F. Brazil, Mercosur and the integration of South America. **Journal of Studies and Research on the Americas,** Vol. 1, No. 1, Aug-Dec, 2007.

WACKERMANN, G. **Les frontières dans un monde en mouvement.** 2 ed. Ellipses, 2003.

ZEBRAL FILHO, S T. B. MARIZ, W. **The New Dynamics of Regional Development in Brazil: Globalization, Socio-Economic Inequalities and Integration.** Available at: http://www.academia.edU/1108887/A NOVA DINAMICA DO DESENVIMENTO REGIONAL NO BRASIL Globalizacao Desigualdades Socio-Economicas e Integracao.

Buy your books fast and straightforward online - at one of world's fastest growing online book stores! Environmentally sound due to Print-on-Demand technologies.

Buy your books online at
www.morebooks.shop

Kaufen Sie Ihre Bücher schnell und unkompliziert online – auf einer der am schnellsten wachsenden Buchhandelsplattformen weltweit! Dank Print-On-Demand umwelt- und ressourcenschonend produzi ert.

Bücher schneller online kaufen
www.morebooks.shop

Printed by Books on Demand GmbH, Norderstedt / Germany